Gran Bretagna: Commissioners of longitude

A Table Of Proportional Logarithms

Gran Bretagna: Commissioners of longitude

A Table Of Proportional Logarithms

ISBN/EAN: 9783741126840

Manufactured in Europe, USA, Canada, Australia, Japa

Cover: Foto ©Thomas Meinert / pixelio.de

Manufactured and distributed by brebook publishing software
(www.brebook.com)

Gran Bretagna: Commissioners of longitude

A Table Of Proportional Logarithms

A

T A B L E

O F

Proportional Logarithms;

To be ufed with the

ASTRONOMICAL AND NAUTICAL

E P. H E M E R I S.

L O N D O N:

Printed by W. Bowyer and J. Nichols:

AND SOLD BY

J. Nourse, Bookfeller to his Majefty, in the Strand;
and Meff. Mount and Page, on Towerhill.

MDCC·LXVI.

PROPORTIONAL

"	0	1	2	3	4	5	6
0		2.2553	1.9542	1.7782	1.6532	1.5563	1.4771
1	4.0334	2.2481	1.9506	1.7757	1.6514	1.5548	1.4759
2	3.7324	2.2410	1.9470	1.7733	1.6496	1.5534	1.4747
3	3.5563	2.2341	1.9435	1.7710	1.6478	1.5520	1.4735
4	3.4313	2.2272	1.9400	1.7686	1.6460	1.5505	1.4723
5	3.3344	2.2205	1.9365	1.7662	1.6442	1.5491	1.4711
6	3.2553	2.2139	1.9331	1.7639	1.6425	1.5477	1.4699
7	3.1883	2.2073	1.9296	1.7616	1.6407	1.5463	1.4687
8	3.1303	2.2009	1.9262	1.7592	1.6390	1.5449	1.4676
9	3.0792	2.1946	1.9228	1.7570	1.6372	1.5435	1.4664
10	3.0334	2.1883	1.9195	1.7546	1.6355	1.5420	1.4652
11	2.9920	2.1821	1.9161	1.7524	1.6337	1.5406	1.4640
12	2.9542	2.1761	1.9128	1.7501	1.6320	1.5393	1.4629
13	2.9195	2.1701	1.9096	1.7478	1.6303	1.5379	1.4617
14	2.8873	2.1642	1.9063	1.7456	1.6286	1.5365	1.4605
15	2.8573	2.1584	1.9031	1.7434	1.6269	1.5351	1.4594
16	2.8293	2.1526	1.8999	1.7411	1.6252	1.5337	1.4582
17	2.8030	2.1469	1.8967	1.7389	1.6235	1.5323	1.4571
18	2.7782	2.1413	1.8935	1.7368	1.6218	1.5310	1.4559
19	2.7546	2.1358	1.8904	1.7345	1.6201	1.5296	1.4548
20	2.7324	2.1303	1.8873	1.7324	1.6184	1.5283	1.4536
21	2.7112	2.1249	1.8842	1.7302	1.6168	1.5269	1.4525
22	2.6910	2.1196	1.8811	1.7281	1.6151	1.5255	1.4513
23	2.6717	2.1143	1.8781	1.7259	1.6134	1.5242	1.4502
24	2.6532	2.1091	1.8751	1.7238	1.6118	1.5229	1.4491
25	2.6355	2.1040	1.8720	1.7216	1.6102	1.5215	1.4479
26	2.6184	2.0989	1.8690	1.7195	1.6085	1.5202	1.4468
27	2.6021	2.0939	1.8661	1.7175	1.6069	1.5189	1.4457
28	2.5862	2.0889	1.8631	1.7153	1.6053	1.5175	1.4446
29	2.5710	2.0840	1.8602	1.7133	1.6037	1.5162	1.4435
30	2.5563	2.0792	1.8573	1.7112	1.6021	1.5149	1.4424

LOGARITHMS.

"	0	1	2	3	4	5	6
31	2.5420	2.0744	1.8544	1.7091	1.6004	1.5136	1.4412
32	2.5283	2.0696	1.8516	1.7071	1.5988	1.5123	1.4401
33	2.5149	2.0649	1.8487	1.7050	1.5973	1.5110	1.4390
34	2.5019	2.0603	1.8459	1.7030	1.5957	1.5097	1.4379
35	2.4893	2.0557	1.8431	1.7010	1.5941	1.5084	1.4368
36	2.4771	2.0512	1.8403	1.6990	1.5925	1.5071	1.4357
37	2.4652	2.0466	1.8375	1.6969	1.5909	1.5058	1.4346
38	2.4536	2.0422	1.8347	1.6949	1.5894	1.5045	1.4335
39	2.4424	2.0378	1.8320	1.6930	1.5878	1.5032	1.4325
40	2.4313	2.0334	1.8293	1.6910	1.5862	1.5019	1.4313
41	2.4206	2.0291	1.8266	1.6890	1.5847	1.5006	1.4303
42	2.4102	2.0248	1.8239	1.6871	1.5832	1.4994	1.4292
43	2.3999	2.0206	1.8212	1.6851	1.5816	1.4981	1.4281
44	2.3899	2.0164	1.8186	1.6832	1.5801	1.4968	1.4270
45	2.3802	2.0122	1.8159	1.6812	1.5786	1.4956	1.4260
46	2.3706	2.0081	1.8133	1.6793	1.5770	1.4943	1.4249
47	2.3613	2.0040	1.8107	1.6774	1.5755	1.4931	1.4238
48	2.3522	2.0000	1.8081	1.6755	1.5740	1.4918	1.4228
49	2.3432	1.9960	1.8055	1.6736	1.5725	1.4906	1.4217
50	2.3344	1.9920	1.8030	1.6717	1.5710	1.4893	1.4206
51	2.3259	1.9881	1.8004	1.6698	1.5695	1.4881	1.4196
52	2.3174	1.9842	1.7979	1.6679	1.5680	1.4869	1.4185
53	2.3091	1.9803	1.7954	1.6660	1.5665	1.4856	1.4175
54	2.3010	1.9765	1.7929	1.6642	1.5651	1.4844	1.4165
55	2.2930	1.9727	1.7904	1.6623	1.5636	1.4832	1.4154
56	2.2852	1.9689	1.7879	1.6605	1.5621	1.4820	1.4143
57	2.2775	1.9652	1.7855	1.6587	1.5607	1.4808	1.4133
58	2.2700	1.9615	1.7830	1.6568	1.5592	1.4795	1.4121
59	2.2626	1.9579	1.7805	1.6550	1.5577	1.4783	1.4112
60	2.2553	1.9542	1.7782	1.6532	1.5561	1.4771	1.4102

PROPORTIONAL

"	7	8	9	10	11	12	13
0	1.4102	1.3522	1.3010	1.2553	1.2139	1.1761	1.1413
1	1.4091	1.3513	1.3002	1.2545	1.2132	1.1755	1.1408
2	1.4081	1.3504	1.2995	1.2538	1.2125	1.1749	1.1402
3	1.4071	1.3495	1.2986	1.2531	1.2119	1.1743	1.1397
4	1.4060	1.3486	1.2978	1.2524	1.2112	1.1737	1.1391
5	1.4050	1.3477	1.2970	1.2517	1.2106	1.1731	1.1385
6	1.4040	1.3468	1.2962	1.2510	1.2099	1.1725	1.1380
7	1.4030	1.3459	1.2954	1.2502	1.2093	1.1719	1.1374
8	1.4020	1.3450	1.2946	1.2495	1.2086	1.1713	1.1369
9	1.4010	1.3441	1.2939	1.2488	1.2080	1.1707	1.1363
10	1.3999	1.3432	1.2931	1.2481	1.2073	1.1701	1.1358
11	1.3989	1.3423	1.2923	1.2474	1.2067	1.1695	1.1352
12	1.3979	1.3415	1.2915	1.2467	1.2061	1.1689	1.1347
13	1.3969	1.3406	1.2907	1.2459	1.2054	1.1683	1.1341
14	1.3959	1.3397	1.2899	1.2452	1.2047	1.1677	1.1336
15	1.3949	1.3388	1.2891	1.2445	1.2041	1.1671	1.1331
16	1.3939	1.3379	1.2883	1.2438	1.2035	1.1665	1.1325
17	1.3929	1.3370	1.2875	1.2431	1.2028	1.1659	1.1319
18	1.3919	1.3362	1.2868	1.2424	1.2022	1.1654	1.1314
19	1.3909	1.3353	1.2860	1.2417	1.2015	1.1648	1.1309
20	1.3899	1.3344	1.2852	1.2410	1.2009	1.1642	1.1303
21	1.3890	1.3336	1.2845	1.2403	1.2003	1.1636	1.1298
22	1.3880	1.3327	1.2837	1.2396	1.1996	1.1630	1.1292
23	1.3870	1.3318	1.2829	1.2389	1.1990	1.1624	1.1287
24	1.3860	1.3310	1.2821	1.2382	1.1984	1.1619	1.1282
25	1.3850	1.3301	1.2814	1.2375	1.1977	1.1613	1.1276
26	1.3841	1.3293	1.2806	1.2368	1.1971	1.1607	1.1271
27	1.3831	1.3284	1.2798	1.2362	1.1965	1.1601	1.1266
28	1.3821	1.3275	1.2791	1.2355	1.1958	1.1595	1.1260
29	1.3812	1.3267	1.2783	1.2348	1.1952	1.1589	1.1255
30	1.3802	1.3259	1.2775	1.2341	1.1946	1.1584	1.1249

LOGARITHMS.

"	7	8	9	10	11	12	13
31	1.3792	1.3250	1.2768	1.2334	1.1939	1.1578	1.1244
32	1.3783	1.3241	1.2760	1.2327	1.1933	1.1572	1.1238
33	1.3773	1.3233	1.2753	1.2320	1.1927	1.1566	1.1233
34	1.3763	1.3224	1.2745	1.2313	1.1920	1.1560	1.1228
35	1.3754	1.3216	1.2737	1.2306	1.1914	1.1555	1.1222
36	1.3745	1.3208	1.2730	1.2300	1.1908	1.1549	1.1217
37	1.3735	1.3199	1.2722	1.2293	1.1902	1.1543	1.1212
38	1.3725	1.3191	1.2715	1.2286	1.1895	1.1537	1.1206
39	1.3716	1.3183	1.2707	1.2279	1.1889	1.1532	1.1201
40	1.3706	1.3174	1.2700	1.2272	1.1883	1.1526	1.1196
41	1.3697	1.3166	1.2692	1.2265	1.1877	1.1520	1.1191
42	1.3688	1.3158	1.2685	1.2259	1.1871	1.1515	1.1186
43	1.3678	1.3149	1.2677	1.2252	1.1864	1.1509	1.1180
44	1.3669	1.3141	1.2670	1.2245	1.1858	1.1503	1.1175
45	1.3660	1.3133	1.2663	1.2239	1.1852	1.1498	1.1170
46	1.3650	1.3124	1.2655	1.2232	1.1846	1.1492	1.1164
47	1.3641	1.3116	1.2648	1.2225	1.1840	1.1486	1.1159
48	1.3632	1.3108	1.2640	1.2218	1.1834	1.1481	1.1154
49	1.3622	1.3099	1.2633	1.2212	1.1828	1.1475	1.1148
50	1.3613	1.3091	1.2626	1.2205	1.1822	1.1469	1.1143
51	1.3604	1.3083	1.2618	1.2198	1.1816	1.1464	1.1138
52	1.3595	1.3075	1.2611	1.2192	1.1809	1.1458	1.1133
53	1.3585	1.3067	1.2603	1.2185	1.1803	1.1452	1.1128
54	1.3576	1.3059	1.2596	1.2178	1.1797	1.1447	1.1123
55	1.3567	1.3050	1.2589	1.2172	1.1791	1.1441	1.1117
56	1.3558	1.3042	1.2582	1.2165	1.1785	1.1435	1.1112
57	1.3549	1.3034	1.2574	1.2159	1.1779	1.1430	1.1107
58	1.3540	1.3026	1.2567	1.2152	1.1773	1.1424	1.1102
59	1.3531	1.3018	1.2560	1.2145	1.1767	1.1419	1.1096
60	1.3522	1.3010	1.2553	1.2130	1.1761	1.1413	1.1091

PROPORTIONAL

″	14′	15′	16′	17′	18′	19′	20′	21′
0	1.1091	1.0792	1.0512	1.0248	0000	9765	9542	9331
1	1.1086	1.0787	1.0507	1.0244	9996	9761	9539	9327
2	1.1081	1.0782	1.0502	1.0240	9992	9757	9535	9323
3	1.1076	1.0777	1.0498	1.0235	9988	9754	9532	9320
4	1.1071	1.0772	1.0493	1.0231	9984	9750	9528	9317
5	1.1066	1.0768	1.0489	1.0227	9980	9746	9524	9313
6	1.1061	1.0763	1.0484	1.0223	9976	9742	9521	9310
7	1.1055	1.0758	1.0480	1.0216	9972	9738	9517	9306
8	1.1050	1.0753	1.0475	1.0214	9968	9735	9513	9303
9	1.1045	1.0749	1.0471	1.0210	9964	9731	9510	9300
10	1.1040	1.0744	1.0466	1.0206	9960	9727	9506	9296
11	1.1035	1.0739	1.0462	1.0201	9956	9723	9503	9293
12	1.1030	1.0734	1.0458	1.0197	9952	9720	9499	9289
13	1.1025	1.0729	1.0453	1.0193	9948	9716	9495	9286
14	1.1020	1.0725	1.0448	1.0189	9944	9712	9492	9282
15	1.1015	1.0720	1.0444	1.0185	9940	9708	9488	9279
16	1.1009	1.0715	1.0440	1.0180	9936	9704	9485	9276
17	1.1004	1.0710	1.0435	1.0176	9932	9701	9481	9272
18	1.0999	1.0706	1.0431	1.0172	9928	9697	9478	9269
19	1.0994	1.0701	1.0426	1.0168	9924	9693	9474	9265
20	1.0989	1.0696	1.0422	1.0164	9920	9689	9470	9262
21	1.0984	1.0692	1.0418	1.0160	9916	9686	9467	9259
22	1.0979	1.0687	1.0413	1.0155	9912	9682	9463	9255
23	1.0974	1.0682	1.0408	1.0151	9908	9678	9460	9252
24	1.0969	1.0678	1.0404	1.0147	9905	9675	9456	9249
25	1.0964	1.0673	1.0400	1.0143	9901	9671	9453	9245
26	1.0959	1.0668	1.0395	1.0139	9897	9667	9449	9242
27	1.0954	1.0663	1.0391	1.0135	9893	9664	9446	9238
28	1.0949	1.0659	1.0386	1.0130	9889	9660	9442	9235
29	1.0944	1.0654	1.0382	1.0126	9885	9656	9439	9231
30	1.0939	1.0649	1.0378	1.0122	9881	9652	9435	9228

″	14′	15′	16′	17′	18′	19′	20′	21′
31	1.0934	1.0645	1.0373	1.0118	.9877	.9648	.9431	.9225
32	1.0929	1.0640	1.0369	1.0114	.9873	.9645	.9428	.9221
33	1.0924	1.0635	1.0365	1.0110	.9869	.9641	.9425	.9218
34	1.0919	1.0631	1.0360	1.0106	.9865	.9637	.9421	.9215
35	1.0914	1.0626	1.0356	1.0102	.9861	.9634	.9417	.9211
36	1.0909	1.0621	1.0352	1.0098	.9858	.9630	.9414	.9208
37	1.0904	1.0617	1.0347	1.0093	.9854	.9626	.9410	.9205
38	1.0899	1.0612	1.0343	1.0089	.9850	.9623	.9407	.9201
39	1.0894	1.0608	1.0339	1.0085	.9846	.9619	.9404	.9198
40	1.0889	1.0603	1.0334	1.0081	.9842	.9615	.9400	.9195
41	1.0884	1.0598	1.0330	1.0077	.9838	.9612	.9396	.9191
42	1.0880	1.0594	1.0326	1.0073	.9834	.9608	.9393	.9188
43	1.0875	1.0589	1.0321	1.0069	.9830	.9604	.9389	.9185
44	1.0870	1.0584	1.0317	1.0065	.9826	.9601	.9386	.9181
45	1.0865	1.0580	1.0313	1.0061	.9823	.9597	.9383	.9178
46	1.0860	1.0575	1.0308	1.0057	.9819	.9593	.9379	.9175
47	1.0855	1.0571	1.0304	1.0053	.9815	.9590	.9375	.9171
48	1.0850	1.0566	1.0300	1.0049	.9811	.9586	.9372	.9168
49	1.0845	1.0561	1.0295	1.0044	.9807	.9582	.9368	.9165
50	1.0840	1.0557	1.0291	1.0040	.9803	.9579	.9365	.9161
51	1.0835	1.0552	1.0287	1.0036	.9800	.9575	.9362	.9158
52	1.0830	1.0548	1.0282	1.0032	.9796	.9571	.9358	.9155
53	1.0826	1.0543	1.0278	1.0028	.9792	.9568	.9355	.9151
54	1.0821	1.0539	1.0274	1.0024	.9788	.9564	.9351	.9148
55	1.0816	1.0534	1.0269	1.0020	.9784	.9560	.9348	.9145
56	1.0811	1.0529	1.0265	1.0016	.9780	.9557	.9344	.9141
57	1.0806	1.0525	1.0261	1.0012	.9777	.9553	.9341	.9138
58	1.0801	1.0520	1.0257	1.0008	.9773	.9549	.9337	.9135
59	1.0796	1.0516	1.0252	1.0004	.9769	.9546	.9334	.9132
60	1.0792	1.051	1.0248	1.0000	.9765	.9542	.9331	.9128

"	22 ′	23 ′	24 ′	25 ′	26 ′	27 ′	28 ′	29 ′	30 ′	31 ′
0	9128	8935	8751	8573	8403	8239	8081	7929	7782	7639
1	9125	8932	8748	8570	8400	8236	8078	7926	7779	7637
2	9122	8929	8745	8567	8397	8234	8076	7924	7776	7634
3	9119	8926	8742	8565	8395	8231	8073	7921	7774	7632
4	9115	8923	8739	8562	8392	8228	8071	7919	7772	7630
5	9112	8920	8736	8559	8389	8225	8068	7916	7769	7627
6	9109	8917	8733	8556	8386	8223	8066	7914	7767	7625
7	9105	8913	8730	8553	8383	8220	8063	7911	7764	7623
8	9102	8910	8727	8550	8381	8217	8060	7909	7762	7620
9	9099	8907	8724	8547	8378	8215	8058	7906	7760	7618
10	9096	8904	8721	8544	8375	8212	8055	7904	7757	7616
11	9092	8901	8718	8541	8372	8209	8053	7901	7755	7613
12	9089	8898	8715	8539	8370	8207	8050	7899	7753	7611
13	9086	8895	8712	8536	8367	8204	8047	7996	7750	7609
14	9082	8891	8709	8533	8364	8202	8045	7994	7748	7606
15	9079	8888	8706	8530	8361	8199	8043	7891	7745	7604
16	9076	8885	8703	8527	8358	8196	8040	7889	7743	7602
17	9073	8882	8700	8524	8356	8194	8037	7886	7740	7599
18	9070	8879	8697	8522	8353	8191	8035	7884	7738	7597
19	9066	8876	8694	8519	8350	8188	8032	7881	7736	7595
20	9063	8873	8691	8516	8347	8186	8030	7879	7733	7592
21	9060	8870	8688	8513	8345	8183	8027	7877	7731	7590
22	9056	8867	8685	8510	8342	8180	8024	7874	7729	7588
23	9053	8864	8682	8507	8339	8178	8022	7872	7726	7586
24	9050	8861	8679	8504	8337	8175	8020	7869	7724	7583
25	9047	8857	8676	8501	8334	8172	8017	7867	7721	7581
26	9044	8854	8673	8498	8331	8170	8014	7864	7719	7579
27	9041	8851	8670	8496	8328	8167	8012	7862	7717	7577
28	9037	8848	8667	8493	8326	8164	8009	7859	7714	7574
29	9034	8845	8664	8490	8323	8162	8007	7857	7712	7572
30	9031	8842	8661	8487	8320	8159	8004	7855	7710	7570

"	22	23	24	25	26	27	28	29	30	31
31	9027	8839	8658	8484	8317	8157	8002	7852	7707	7567
32	9024	8836	8655	8481	8315	8154	7999	7849	7705	7565
33	9021	8833	8652	8479	8312	8152	7997	7847	7703	7563
34	9018	8830	8649	8476	8309	8149	7994	7844	7700	7560
35	9015	8827	8646	8473	8306	8146	7991	7842	7698	7558
36	9012	8824	8643	8470	8304	8144	7989	7840	7696	7556
37	9008	8820	8640	8467	8301	8141	7986	7837	7693	7553
38	9005	8817	8637	8464	8298	8138	7984	7835	7691	7551
39	9002	8814	8635	8462	8296	8136	7981	7832	7688	7549
40	8999	8811	8632	8459	8293	8133	7979	7830	7686	7546
41	8995	8808	8629	8456	8290	8130	7976	7827	7683	7544
42	8992	8805	8626	8453	8288	8128	7974	7825	7681	7542
43	8989	8802	8623	8450	8285	8125	7971	7823	7679	7540
44	8986	8799	8620	8448	8282	8122	7969	7820	7676	7537
45	8983	8796	8617	8445	8279	8120	7966	7818	7674	7535
46	8980	8793	8614	8442	8277	8117	7964	7815	7672	7533
47	8976	8790	8611	8439	8274	8115	7961	7813	7669	7531
48	8973	8787	8608	8437	8271	8112	7959	7811	7667	7528
49	8970	8784	8605	8434	8268	8109	7956	7808	7665	7526
50	8967	8781	8602	8431	8266	8107	7954	7805	7662	7524
51	8964	8778	8599	8428	8263	8104	7951	7803	7660	7522
52	8960	8775	8596	8425	8260	8102	7949	7801	7658	7519
53	8957	8772	8593	8422	8258	8099	7946	7798	7655	7517
54	8954	8769	8591	8420	8255	8097	7944	7796	7653	7515
55	8951	8766	8588	8417	8252	8094	7941	7793	7651	7512
56	8948	8763	8585	8414	8250	8091	7939	7791	7648	7510
57	8945	8760	8582	8411	8247	8089	7936	7789	7646	7508
58	8942	8757	8579	8408	8244	8086	7934	7786	7644	7506
59	8938	8754	8576	8406	8242	8084	7931	7784	7641	7503
60	8935	8751	8573	8403	8239	8081	7929	7782	7639	7501

C

"	32	33	34	35	36	37	38	39	40	41
0	7501	7368	7238	7112	6990	6871	6755	6642	6532	6425
1	7499	7365	7236	7110	6988	6869	6753	6640	6530	6423
2	7496	7363	7234	7108	6986	6867	6751	6638	6528	6421
3	7494	7361	7232	7106	6984	6865	6749	6637	6527	6420
4	7492	7359	7229	7104	6982	6863	6747	6635	6525	6418
5	7490	7356	7227	7102	6980	6861	6745	6633	6523	6416
6	7488	7354	7225	7100	6978	6859	6743	6631	6521	6414
7	7485	7352	7223	7097	6976	6857	6741	6629	6519	6412
8	7483	7350	7221	7095	6974	6855	6739	6627	6517	6411
9	7481	7348	7219	7093	6972	6853	6738	6625	6516	6409
10	7478	7345	7216	7091	6970	6851	6736	6623	6514	6407
11	7476	7343	7214	7089	6968	6849	6734	6621	6512	6405
12	7474	7341	7212	7087	6966	6847	6732	6620	6510	6404
13	7472	7339	7210	7085	6964	6845	6730	6618	6508	6402
14	7469	7337	7208	7083	6962	6843	6728	6616	6507	6400
15	7467	7335	7206	7081	6960	6841	6726	6614	6505	6398
16	7465	7332	7204	7079	6958	6839	6724	6612	6503	6397
17	7463	7330	7202	7077	6956	6837	6722	6610	6501	6395
18	7461	7328	7200	7075	6954	6836	6721	6609	6500	6393
19	7458	7326	7197	7073	6952	6834	6719	6607	6498	6391
20	7456	7324	7195	7071	6950	6832	6717	6605	6496	6390
21	7454	7322	7193	7069	6948	6830	6715	6603	6494	6388
22	7452	7319	7191	7067	6946	6828	6713	6601	6492	6386
23	7449	7317	7189	7065	6944	6826	6711	6599	6490	6384
24	7447	7315	7187	7063	6942	6824	6709	6598	6489	6383
25	7445	7313	7185	7061	6940	6822	6707	6596	6487	6381
26	7443	7311	7183	7059	6938	6820	6705	6594	6485	6379
27	7441	7309	7181	7057	6936	6818	6704	6592	6484	6377
28	7438	7306	7179	7054	6934	6816	6702	6590	6482	6376
29	7436	7304	7177	7052	6932	6814	6700	6588	6480	6374
30	7434	7302	7175	7050	6930	6812	6698	6587	6478	6372

"	32	33	34	35	36	37	38	39	40	41
31	7431	7300	7172	7048	6928	6810	6696	6585	6476	6370
32	7429	7298	7170	7046	6926	6808	6694	6583	6474	6369
33	7427	7296	7168	7044	6924	6807	6692	6581	6473	6367
34	7425	7293	7166	7042	6922	6805	6690	6579	6471	6465
35	7423	7291	7164	7040	6920	6803	6689	6577	6469	6363
36	7421	7289	7162	7038	6918	6801	6687	6576	6467	6362
37	7418	7287	7160	7036	6916	6799	6685	6574	6465	6360
38	7416	7285	7158	7034	6914	6797	6683	6572	6464	6358
39	7414	7283	7156	7032	6912	6795	6681	6570	6462	6357
40	7411	7281	7153	7030	6910	6793	6679	6568	6460	6355
41	7409	7278	7151	7028	6908	6791	6677	6566	6458	6353
42	7407	7276	7149	7026	6906	6789	6676	6565	6457	6351
43	7405	7274	7147	7024	6904	6787	6674	6563	6455	6349
44	7403	7272	7145	7022	6902	6785	6672	6561	6453	6348
45	7401	7270	7143	7020	6900	6784	6670	6559	6451	6346
46	7398	7268	7141	7018	6898	6782	6668	6557	6449	6344
47	7396	7266	7139	7016	6896	6780	6666	6556	6448	6342
48	7394	7264	7137	7014	6894	6778	6664	6554	6446	6341
49	7392	7261	7135	7912	6892	6776	6662	6552	9444	6339
50	7389	7259	7133	7010	6890	6774	6660	6550	6442	6337
51	7387	7257	7131	7008	6888	6772	6659	6548	6441	6336
52	7385	7255	7128	7006	6886	6770	6657	6546	6439	6334
53	7383	7253	7126	7004	6884	6768	6655	6545	6437	6332
54	7381	7251	7124	7002	6882	6766	6653	6543	6435	6431
55	7378	7248	7122	7000	6880	6764	6651	6541	6434	6329
56	7376	7246	7120	6998	6878	6762	6649	6539	6432	6327
57	7374	7244	7118	6996	6877	6761	6648	6538	6430	6325
58	7372	7242	7116	6994	6875	6759	6646	6536	6428	6323
59	7370	7240	7114	6992	6873	6757	6644	6534	6426	6322
60	7368	7238	7112	6990	6871	6755	6642	6532	6425	6320

"	42′	43′	44′	45′	46′	47′	48′	49′	50′	51′
0	6320	6218	6118	6021	5925	5832	5740	5651	5563	5477
1	6318	6216	6116	6019	5923	5830	5739	5649	5561	5475
2	6317	6214	6115	6017	5912	5838	5737	5648	5560	5474
3	6315	6213	6113	6016	5920	5827	5736	5646	5559	5473
4	6313	6211	6111	6014	5919	5825	5734	5645	5557	5471
5	6311	6209	6110	6012	5917	5824	5733	5643	5556	5470
6	6310	6208	6108	6011	5916	5823	5731	5642	5554	5469
7	6308	6206	6106	6009	5414	5821	5730	5640	5553	5467
8	6306	6204	6105	6008	5912	5819	5728	5639	5551	5465
9	6305	6203	6103	6006	5911	5818	5727	5637	5550	5464
10	6303	6201	6102	6004	5909	5816	5725	5636	5548	5463
11	6301	6199	6100	6003	5908	5815	5724	5634	5547	5461
12	6300	6198	6099	6001	5906	5813	5722	5633	5546	5460
13	6298	6196	6097	6000	5905	5812	5721	5631	5544	5458
14	6296	6194	6095	5998	5903	5810	5719	5630	5543	5457
15	6294	6193	6094	5997	5902	5809	5718	5629	5541	5456
16	6293	6191	6092	5995	5900	5807	5716	5627	5540	5454
17	6291	6189	6090	5993	5898	5805	5715	5626	5538	5453
18	6289	6188	6089	5992	5897	5804	5713	5624	5537	5452
19	6287	6186	6087	5990	5895	5802	5712	5623	5535	5450
20	6286	6184	6085	5988	5894	5801	5710	5621	5534	5449
21	6284	6183	6084	5987	5892	5800	5709	5620	5533	5447
22	6282	6181	6082	5985	5890	5798	5707	5618	5531	5446
23	6281	6179	6080	5984	5889	5796	5706	5617	5530	5444
24	6279	6178	6079	5982	5888	5795	5704	5615	5528	5443
25	6277	6176	6077	5980	5886	5793	5703	5614	5527	5441
26	6275	6174	6075	5979	5884	5792	5701	5612	5525	5440
27	6274	6173	6074	5977	5883	5790	5700	5611	5524	5439
28	6272	6171	6072	5976	5881	5789	5698	5609	5522	5437
29	6270	6169	6071	5974	5882	5787	5697	5602	5521	5436
30	6268	6168	6069	5972	5878	5786	5695	5607	5520	5435

"	42	43	44	45	46	47	48	49	50	51
31	6267	6166	6067	5971	5876	5784	5694	5605	5518	5433
32	6265	6164	6066	5969	5875	5783	5693	5604	5517	5432
33	6264	6163	6064	5968	5874	5781	5691	5602	5516	5430
34	6262	6161	6062	5966	5872	5779	5689	5601	5514	5429
35	6260	6159	6061	5964	5870	5778	5688	5599	5512	5727
36	6259	6158	6059	5963	5869	5777	5686	5598	5511	5426
37	6257	6156	6058	5961	5867	5775	5685	5596	5510	5425
38	6255	6154	6056	5960	5866	5773	5683	5595	5508	5423
39	6254	6153	6055	5958	5864	5772	5682	5594	5507	5422
40	6252	6151	6053	5957	5862	5770	5680	5592	5505	5420
41	6250	6149	6051	5955	5861	5769	5679	5590	5504	5419
42	6248	6148	6050	5954	5860	5768	5677	5589	5503	5418
43	6247	6146	6048	5952	5858	5766	5676	5587	5501	5416
44	6245	6144	6046	5950	5856	5764	5674	5586	5500	5415
45	6243	6143	6045	5949	5855	5763	5673	5585	5498	5414
46	6241	6141	6043	5947	5853	5761	5671	5583	5497	5412
47	6240	6139	6041	5945	5852	5760	5670	5582	5495	5411
48	6238	6138	6040	5944	5850	5758	5669	5580	5494	5409
49	6236	6136	6038	5942	5849	5757	5667	5579	5492	5408
50	6235	6134	6037	5941	5847	5755	5665	5577	5491	5406
51	6233	6133	6035	5939	5846	5754	5664	5576	5490	5405
52	6231	6131	6033	5938	5844	5752	5662	5574	5488	5404
53	6230	6130	6032	5936	5842	5751	5661	5573	5487	5402
54	6228	6128	6030	5935	5841	5749	5660	5572	5486	5401
55	6226	6126	6028	5933	5839	5748	5658	5570	5484	5399
56	6225	6125	6027	5931	5838	5746	5656	5569	5482	5398
57	6223	6123	6025	5930	5836	5745	5655	5567	5481	5397
58	6221	6121	6024	5928	5835	5743	5654	5566	5480	5395
59	6220	6120	6022	5927	5833	5742	5652	5564	5478	5394
60	6218	6118	6021	5925	5832	5740	5651	5563	5477	5393

"	52	53	54	55	56	57	58	59	h. 0 t. 0	h. 1 t. 1
0	5395	5310	5229	5149	5071	4994	4918	4844	4771	4699
1	5391	5308	5227	5148	5069	4992	4917	4843	4770	4698
2	5390	5307	5226	5146	5068	4991	4916	4842	4769	4697
3	5389	5306	5225	5145	5067	4990	4915	4841	4768	4696
4	5387	5304	5223	5144	5065	4989	4913	4839	4766	4694
5	5386	5303	5222	5142	5064	4987	4912	4838	4765	4693
6	5384	5302	5221	5141	5063	4986	4911	4837	4764	4692
7	5383	5300	5219	5140	5062	4985	4910	4835	4763	4691
8	5381	5299	5218	5138	5060	4984	4908	4834	4761	4690
9	5380	5298	5217	5137	5059	4983	4907	4833	4760	4689
10	5379	5296	5215	5136	5058	4981	4906	4832	4759	4687
11	5377	5295	5214	5134	5056	4980	4905	4831	4758	4686
12	5376	5294	5215	5133	5055	4979	4903	4830	4757	4685
13	5374	5292	5211	5132	5054	4977	4902	4828	4755	4684
14	5373	5291	5210	5130	5053	4976	4901	4827	4754	4683
15	5372	5290	5209	5129	5051	4975	4900	4826	4753	4682
16	5370	5288	5207	5128	5050	4973	4898	4824	4752	4680
17	5369	5287	5206	5127	5049	4972	4897	4823	4751	4679
18	5368	5285	5205	5125	5048	4971	4896	4822	4750	4678
19	5366	5284	5203	5124	5046	4970	4895	4821	4748	4677
20	5365	5283	5202	5123	5045	4968	4893	4820	4747	4676
21	5364	5281	5201	5122	5044	4967	4892	4819	4746	4675
22	5362	5280	5199	5120	5042	4966	4891	4817	4745	4673
23	5361	5278	5198	5119	5041	4965	4890	4816	4743	4672
24	5359	5277	5197	5118	5040	4964	4889	4815	4742	4671
25	5358	5276	5195	5116	5038	4962	4887	4813	4741	4670
26	5356	5274	5194	5115	5037	4961	4886	4812	4740	4669
27	5355	5273	5193	5114	5036	4960	4885	4811	4739	4668
28	5354	5272	5191	5112	5735	4958	4883	4810	4737	4666
29	5353	5270	5190	5111	5033	4957	4882	4809	4736	4665
30	5351	5269	5189	5110	5032	4956	4881	4808	4735	4664

LOGARITHMS.

"	52	53	54	55	56	57	58	59	h. 0 / 1. 0	h. 0 / 1. 1
31	5350	5268	5187	5108	5031	4955	4880	4806	4734	4663
32	5348	5266	5186	5107	5029	4953	4878	4805	4733	4661
33	5347	5265	5185	5106	5028	4952	4877	4804	4732	4660
34	5345	5264	5183	5104	5027	4951	4876	4802	4730	4659
35	5344	5262	5182	5103	5026	4950	4875	4801	4729	4658
36	5343	5261	5181	5102	5025	4949	4874	4800	4728	4657
37	5341	5260	5179	5100	5023	4947	4872	4799	4727	4656
38	5340	5258	5178	5099	5022	4946	4871	4798	4725	4654
39	5339	5257	5177	5098	5021	4945	4870	4797	4724	4653
40	5337	5255	5175	5097	5019	4943	4869	4795	4723	4652
41	5336	5254	5174	5095	5018	4942	4867	4794	4722	4651
42	5335	5253	5173	5094	5017	4941	4866	4793	4721	4650
43	5333	5251	5171	5093	5015	4940	4865	4792	4719	4648
44	5332	5250	5170	5091	5014	4938	4864	4790	4718	4647
45	5331	5249	5169	5090	5013	4937	4863	4789	4717	4646
46	5329	5247	5167	5089	5012	4936	4861	4788	4716	4645
47	5328	5246	5166	5087	5010	4934	4860	4787	4715	4644
48	5326	5245	5165	5086	5009	4933	4859	4786	4714	4643
49	5325	5243	5163	5085	5008	4932	4858	4784	4712	4641
50	5323	5242	5162	5084	5006	4931	4856	4783	4711	4640
51	5322	5241	5161	5082	5005	4930	4855	4782	4710	4639
52	5321	5239	5159	5081	5004	4928	4854	4781	4709	4638
53	5319	5238	5158	5080	5003	4927	4853	4779	4708	4637
54	5318	5237	5157	5079	5002	4926	4852	4778	4707	4636
55	5317	5235	5155	5077	5000	4924	4850	4777	4705	4634
56	5315	5234	5154	5076	4999	4923	4849	4776	4704	4633
57	5314	5233	5153	5075	4998	4922	4848	4775	4703	4632
58	5312	5231	5152	5073	4996	4921	4846	4773	4702	4631
59	5311	5230	5150	5072	4995	4919	4845	4772	4700	4630
60	5310	5229	5149	5071	4994	4918	4844	4771	4699	4620

"	h. / 1. 2	h. / 1. 3	h. / 1. 4	h. / 1. 5	h. / 1. 6	h. / 1. 7	h. / 1. 8	h. / 1. 9	h. / 1. 10	h. / 1. 11
0	4629	4559	4491	4424	4357	4292	4228	4164	4102	4040
1	4627	4558	4490	4422	4356	4291	4226	4163	4101	4039
2	4626	4557	4489	4421	4355	4290	4225	4162	4100	4038
3	4625	4556	4488	4420	4354	4289	4224	4161	4099	4037
4	4624	4555	4486	4419	4353	4287	4223	4160	4098	4036
5	4623	4553	4485	4418	4352	4286	4222	4159	4097	4035
6	4622	4552	4484	4417	4351	4285	4221	4158	4096	4034
7	4620	4551	4483	4416	4349	4284	4220	4157	4094	4033
8	4619	4550	4482	4415	4348	4283	4219	4156	4093	4032
9	4618	4549	4481	4414	4247	4282	4218	4155	4092	4031
10	4617	4548	4479	4412	4346	4281	4217	4154	4091	4030
11	4616	4547	4478	4411	4345	4280	4216	4153	4090	4029
12	4615	4546	4477	4410	4344	4279	4215	4152	4089	4028
13	4613	4544	4476	4409	4343	4278	4214	4151	4088	4027
14	4612	4543	4475	4408	4342	4277	4213	4150	4087	4026
15	4611	4542	4474	4407	4341	4276	4212	4149	4086	4025
16	4610	4541	4473	4406	4340	4275	4211	4147	4085	4024
17	4609	4540	4472	4405	4339	4274	4210	4146	4084	4023
18	4608	4539	4471	4404	4338	4273	4209	4145	4083	4022
19	4606	4537	4469	4402	4336	4271	4207	4144	4082	4021
20	4605	4536	4468	4401	4335	4270	4206	4143	4081	4020
21	4604	4535	4467	4400	4334	4269	4205	4142	4080	4019
22	4603	4534	4466	4399	4333	4268	4204	4141	4079	4018
23	4602	4533	4465	4398	4332	4267	4203	4140	4078	4017
24	4601	4532	4464	4397	4331	4266	4202	4139	4077	4016
25	4600	4530	4463	4396	4330	4265	4201	4138	4076	4015
26	4598	4529	4461	4395	4329	4264	4200	4137	4075	4014
27	4597	4528	4460	4394	4328	4263	4199	4136	4074	4013
28	4596	4527	4459	4392	4327	4262	4198	4135	4073	4012
29	4595	4526	4458	4391	4326	4261	4197	4134	4072	4011
30	4594	4525	4457	4390	4325	4260	4196	4133	4071	4010

LOGARITHMS.

	h. '	h. '	h. '	h. '	h. '	h. '	h. '	h. '	h. '	h. '
"	1. 2	1. 3	1. 4	1. 5	1. 6	1. 7	1. 8	1. 9	1. 10	1. 11
31	4593	4524	4456	4389	4323	4258	4195	4132	4070	4009
32	4591	4523	4455	4388	4322	4257	4194	4131	4069	4008
33	4590	4522	4454	4387	4321	4256	4193	4130	4068	4007
34	4589	4520	4452	4386	4320	4255	4191	4129	4067	4006
35	4588	4519	4451	4385	4319	4254	4190	4128	4066	4005
36	4587	4518	4450	4384	4318	4253	4189	4127	4065	4004
37	4586	4517	4449	4382	4317	4252	4188	4126	4064	4003
38	4585	4516	4448	4381	4316	4251	4187	4125	4063	4002
39	4584	4515	4447	4380	4315	4250	4186	4124	4062	4001
40	4582	4513	4446	4379	4313	4249	4185	4122	4061	4000
41	4581	4512	4445	4378	4312	4248	4184	4121	4060	3999
42	4580	4511	4414	4377	4311	4247	418.	4120	4059	3998
43	4579	4510	4442	4376	4310	4246	4182	4119	4057	3997
44	4578	4509	4441	4375	4309	4245	4181	4118	4056	3996
45	4577	4508	4440	4374	4308	4244	4180	4117	4055	3995
46	4575	4507	4439	4372	4307	4242	4179	4116	4054	3993
47	4574	4506	4438	4371	4306	4241	4178	4115	4053	3992
48	4573	4505	4437	4370	4305	4240	4177	4114	4052	3991
49	4572	4503	4436	4369	4304	4239	4176	4113	4051	3990
50	4571	4502	4435	4368	4303	4238	4175	4112	4050	3989
51	4570	4501	4434	4367	4302	4237	4174	4111	4049	3988
52	4568	4500	4432	4366	4300	4236	4173	4110	4048	3987
53	4567	4499	4431	4365	4299	4235	4172	4109	4047	3986
54	4566	4498	4430	4364	4298	4234	4171	4108	4046	3985
55	4565	4496	4429	4363	4297	4233	4169	4107	4045	3984
56	4564	4595	4428	4362	4296	4232	4168	4106	4044	3983
57	4563	4494	4427	4361	4295	4231	4167	4105	4043	3982
58	4561	4493	4426	4359	4294	4230	4166	4104	4042	3981
59	4560	4492	4425	4358	4293	4229	4165	4103	4041	3980
60	4559	4491	4424	4357	4292	4228	4164	4102	4040	3979

D

"	h. ' 1. 12	h. ' 1. 13	h. ' 1. 14	h. ' 1. 15	h. ' 1. 16	h. ' 1. 17	h. ' 1. 18	h. ' 1. 19	h. ' 1. 20	h. ' 1. 21
0	3979	3919	3860	3802	3745	3688	3632	3576	3522	3468
1	3978	3918	3859	3801	3744	3687	3631	3575	3521	3467
2	3977	3917	3858	3800	3743	3686	3630	3574	3520	3466
3	3976	3917	3857	3799	3742	3685	3629	3574	3519	3465
4	3975	3916	3856	3798	3741	3684	3628	3573	3518	3464
5	3974	3915	3855	3797	3740	3683	3627	3572	3517	3463
6	3973	3914	3855	3796	3739	3682	3626	3571	3516	3463
7	3972	3913	3854	3795	3738	3681	3625	3570	3515	3462
8	3971	3912	3853	3794	3737	3680	3624	3569	3514	3461
9	3970	3911	3852	3793	3736	3679	3623	3568	3514	3460
10	3969	3910	3851	3792	3735	3678	3622	3567	3513	3459
11	3968	3909	3850	3791	3734	3677	3621	3566	3512	3458
12	3967	3908	3849	3791	3733	3677	3621	3565	3511	3457
13	3966	3907	3848	3790	3732	3676	3620	3564	3510	3456
14	3965	3906	3847	3789	3731	3675	3619	3563	3509	3455
15	3964	3905	3846	3788	3730	3674	3618	3563	3508	3454
16	3963	3904	3845	3787	3729	3673	3617	3562	3507	3454
17	3962	3903	3844	3786	3728	3672	3616	3561	3506	3453
18	3961	3902	3843	3785	3727	3671	3615	3560	3506	3452
19	3960	3901	3842	3784	3726	3670	3614	3559	3505	3451
20	3959	3900	3841	3783	3725	3669	3613	3558	3504	3450
21	3958	3899	3840	3782	3725	3668	3612	3557	3503	3449
22	3957	3898	3839	3781	3724	3667	3611	3556	3502	3448
23	3956	3897	3838	3780	3723	3666	3610	3555	3501	3447
24	3955	3896	3837	3779	3722	3665	3610	3555	3500	3446
25	3954	3895	3836	3778	3721	3664	3609	3554	3499	3445
26	3953	3894	3835	3777	3720	3663	3608	3553	3498	3445
27	3952	3893	3834	3776	3719	3663	3607	3552	3497	3444
28	3951	3892	3833	3775	3718	3662	3606	3551	3496	3443
29	3950	3891	3832	3774	3717	3661	3605	3550	3496	3442
30	3949	3890	3831	3773	3716	3660	3604	3549	3495	3441

LOGARITHMS.

"	h. 1.12	h. 1.13	h. 1.14	h. 1.15	h. 1.16	h. 1.17	h. 1.18	h. 1.19	h. 1.20	h. 1.21
31	3948	3889	3830	3772	3715	3659	3603	3548	3494	3440
32	3947	3888	3829	3771	3714	3658	3602	3547	3493	3439
33	3946	3887	3828	3770	3713	3657	3601	3546	3492	3438
34	3945	3886	3827	3769	3712	3656	3600	3545	3491	3436
35	3944	3885	3826	3768	3711	3655	3599	3544	3490	3437
36	3943	3884	3825	3768	3710	3654	3598	3544	3489	3436
37	3942	3883	3824	3767	3709	3653	3597	3543	3488	3435
38	3941	3882	3823	3766	3708	3652	3596	3542	3487	3434
39	3940	3881	3822	3765	3708	3651	3596	3541	3487	3433
40	3939	3880	3821	3764	3707	3650	3595	3540	3486	3432
41	3938	3879	3820	3763	3706	3649	3594	3539	3485	3431
42	3937	3878	3820	3762	3705	3649	3593	3538	3484	3431
43	3936	3877	3819	3761	3704	3648	3592	3537	3483	3430
44	3935	3876	3818	3760	3703	3647	3591	3536	3482	3429
45	3934	3875	3817	3759	3702	3646	3590	3535	3481	3428
46	3933	3874	3816	3758	3701	3645	3589	3534	3480	3427
47	3932	3873	3815	3757	3700	3644	3588	3533	3479	3426
48	3931	3872	3814	3756	3699	3643	3587	3533	3479	3425
49	3930	3871	3813	3755	3698	3642	3586	3532	3478	3424
50	3929	3870	3812	3754	3697	3641	3585	3531	3477	3423
51	3928	3869	3811	3753	3696	3640	3585	3530	3476	3423
52	3927	3868	3810	3752	3695	3639	3584	3529	3475	3422
53	3926	3867	3809	3751	3694	3638	3583	3528	3474	3421
54	3925	3866	3808	3750	3693	3637	3582	3527	3473	3420
55	3924	3865	3807	3749	3692	3636	3581	3526	3472	3419
56	3923	3864	3806	3748	3691	3635	3580	3525	3471	3418
57	3922	3863	3805	3747	3691	3635	3579	3525	3471	3417
58	3921	3862	3804	3746	3690	3634	3578	3524	3470	3416
59	3920	3861	3803	3745	3689	3633	3577	3523	3469	3415
60	3919	3860	3801	3745	3688	3633	3576	3522	3468	3415

PROPORTIONAL

"	h. / 1.22	h. / 1.23	h. / 1.24	h. / 1.25	h. / 1.26	h. / 1.27	h. / 1.28	h. / 1.29	h. / 1.30	h. / 1.31
0	3415	3362	3310	3259	3208	3158	3108	3059	3010	2962
1	3414	3361	3309	3258	3207	3157	3107	3058	3009	2961
2	3413	3360	3308	3257	3206	3156	3106	3057	3009	2961
3	3412	3359	3307	3256	3205	3155	3105	3052	3008	2960
4	3411	3358	3306	3255	3204	3154	3105	3056	3007	2959
5	3410	3358	3306	3254	3203	3153	3104	3055	3006	2958
6	3409	3357	3305	3253	3203	3153	3103	3054	3005	2958
7	3408	3356	3304	3253	3202	3152	3102	3053	3005	2957
8	3407	3355	3303	3252	3201	3151	3101	3052	3004	2956
9	3407	3354	3302	3251	3200	3150	3101	3052	3003	2955
10	3406	3353	3301	3250	3199	3149	3100	3051	3002	2953
11	3405	3352	3300	3249	3198	3148	3099	5050	3001	2954
12	3404	3351	3300	3248	3198	3148	3098	3049	3001	2953
13	3403	3351	3299	3247	3197	3147	3097	3048	3000	2952
14	3402	3350	3298	3247	3196	3146	3090	3047	2999	2951
15	3401	3349	3297	3246	3195	3145	3096	3047	2998	2950
16	3400	3348	3296	3245	3194	3144	3095	3046	2997	2950
17	3400	3347	3295	3244	3193	3143	3094	3045	2997	2949
18	3399	3346	3294	3243	3193	3143	3093	3044	2996	2948
19	3398	3345	3294	3242	3192	3142	3092	3043	2995	2947
20	3397	3344	3293	3241	3191	3141	3091	3043	2994	2946
21	3396	3344	3292	3241	3192	3140	3091	3042	2993	2946
22	3395	3343	3291	3240	3189	3139	3090	3041	2993	2945
23	3394	3342	3290	3239	3188	3138	3089	3040	2992	2944
24	3393	3341	3289	3238	3188	3136	3088	3039	1991	2943
25	3393	3340	3288	3237	3187	3137	3087	3038	2990	2942
26	3392	3339	3287	3236	3186	3136	3686	3038	2989	2942
27	3391	3338	3287	3236	3185	3135	3086	3037	2989	2941
28	3390	3338	3286	3235	3184	3134	3085	3036	2988	2940
29	3389	3337	3285	3234	3183	3133	3084	3035	2987	2939
30	3388	3336	3284	3233	3183	3133	3083	3034	2986	2938

LOGARITHMS.

"	h. 1.22	h. 1.23	h 1.24	h 1.25	h. 1.26	h. 1.27	h 1.28	h. 1.29	h. 1.30	h. 1.31
31	3387	3335	3283	3232	3182	3132	3082	3034	2985	2936
32	3386	3334	3282	3231	3181	3131	3082	3033	2985	2937
33	3386	3333	3282	3231	3180	3130	3081	3032	2984	2936
34	3385	3332	3281	3230	3179	3129	3080	3031	2983	2935
35	3384	3331	3280	3229	3178	3128	3079	3030	2982	2934
36	3383	3331	3279	3228	3178	3128	3078	3030	2981	2935
37	3382	3330	3278	3227	3177	3127	3078	3029	2981	2933
38	3381	3329	3277	3226	3176	3126	3077	3028	2980	2933
39	3380	3328	3276	3225	3175	3125	3076	3027	2979	2931
40	3379	3327	3276	3225	3174	3124	3075	3026	2978	2931
41	3378	3326	3275	3224	3173	3123	3074	3026	2977	2930
42	3378	3325	3274	3223	3173	3123	3073	3025	2977	2920
43	3377	3325	3273	3222	3172	3122	3073	3024	2976	2928
44	3376	3324	3272	3221	3171	3121	3072	3023	2975	2927
45	3375	3323	3271	3220	3170	3120	3071	3022	2974	2927
46	3374	3322	3270	3219	3169	3119	3070	3022	2973	2926
47	3373	3321	3270	3219	3168	3119	3069	3021	2973	2925
48	3372	3320	3269	3218	3168	3118	3069	3020	2972	2924
49	3371	3319	3268	3217	3167	3117	3068	3019	2971	2923
50	3371	3318	3267	3216	3166	3116	3067	3018	2970	2923
51	3370	3318	3266	3215	3165	3115	3066	3018	2969	2922
52	3369	3317	2365	3214	3164	3114	3065	3017	2969	2921
53	3368	3316	3264	3214	3163	3114	3064	3016	2968	2920
54	3367	3315	3264	3213	3163	3113	3064	3015	2967	2920
55	3366	3314	3263	3212	3162	3112	3063	3014	2966	2919
56	3365	3313	3262	3211	3161	3111	3062	3013	2965	2918
57	3365	3313	3261	3210	3160	3110	3061	3013	2965	2917
58	3364	3312	3260	3209	3159	3109	3060	3012	2964	2916
59	3363	3311	3259	3209	3158	3109	3060	3011	2963	2916
60	3362	3310	3259	1206	3158	3108	3059	3010	2902	2915

PROPORTIONAL

"	h. ' 1. 32	h. ' 1. 33	h. ' 1. 34	h. ' 1. 35	h. ' 1. 36	h. ' 1. 37	h. ' 1. 38	h. ' 1. 39	h. ' 1. 40	h. ' 1. 41
0	2915	2868	2821	2775	2730	2685	2640	2596	2553	2510
1	2914	2867	2821	2775	2729	2684	2640	2596	2552	2509
2	2913	2866	2820	2774	2728	2683	2639	2595	2551	2508
3	2912	2866	2819	2773	2728	2683	2638	2594	2551	2507
4	2912	2865	2818	2772	2727	2682	2637	2593	2550	2507
5	2911	2864	2818	2772	2726	2681	2637	2593	2549	2506
6	2910	2863	2817	2771	2725	2681	2636	2592	2548	2505
7	2909	2862	2816	2770	2725	2680	2635	2591	2548	2504
8	2908	2862	2815	2769	2724	2679	2634	2590	2547	2504
9	2908	2861	2815	2769	2723	2678	2634	2590	2546	2503
10	2907	2860	2814	2768	2722	2678	2633	2589	2545	2502
11	2906	2859	2813	2767	2722	2677	2632	2588	2545	2502
12	2905	2859	2812	2766	2721	2676	2632	2588	2544	2501
13	2905	2858	2811	2766	2720	2675	2631	2587	2543	2500
14	2904	2857	2811	2765	2719	2675	2630	2586	2543	2499
15	2903	2856	2810	2764	2719	2674	2629	2585	2542	2499
16	2902	2855	2809	2763	2718	2673	2629	2585	2541	2498
17	2901	2855	2808	2762	2717	2672	2628	2584	2540	2497
18	2901	2854	2808	2762	2716	2672	2627	2583	2540	2497
19	2900	2853	2807	2761	2716	2671	2626	2582	2539	2496
20	2899	2852	2806	2760	2715	2670	2626	2582	2538	2495
21	2898	2852	2805	2760	2714	2669	2625	2581	2538	2494
22	2898	2851	2804	2759	2713	2669	2624	2580	2537	2494
23	2897	2850	2804	2758	2713	2668	2623	2580	2536	2493
24	2896	2849	2803	2757	2712	2667	2623	2579	2535	2492
25	2895	2848	2802	2756	2711	2666	2622	2578	2535	2492
26	2894	2848	2801	2756	2710	2666	2621	2577	2534	2491
27	2894	2847	2801	2755	2710	2665	2621	2577	2533	2490
28	2893	2846	2800	2754	2709	2664	2620	2576	2532	2489
29	2892	2845	2799	2753	2708	2663	2619	2575	2532	2489
30	2891	2845	2798	2753	2707	2663	2618	2574	2531	2488

"	1.32	1.33	1.34	1.35	1.36	1.37	1.38	1.39	1.40	1.41
31	2890	2844	2798	2752	2707	2662	2618	2574	2530	2487
32	2890	2843	2797	2751	2706	2661	2617	2573	2530	2487
33	2889	2842	2796	2750	2705	2660	2616	2572	2529	2486
34	2888	2841	2795	2750	2704	2660	2615	2572	2528	2485
35	2887	2841	2795	2749	2704	2659	2615	2571	2527	2484
36	2887	2840	2794	2748	2703	2658	2614	2570	2527	2484
37	2886	2839	2793	2747	2702	2657	2613	2569	2526	2483
38	2885	2838	2792	2747	2701	2657	2612	2569	2525	2482
39	2884	2838	2792	2746	2701	2656	2612	2568	2525	2482
40	2883	2837	2791	2745	2700	2655	2611	2567	2524	2481
41	2883	2836	2790	2744	2699	2654	2610	2566	2523	2480
42	2882	2835	2789	2744	2698	2654	2610	2566	2522	2480
43	2881	2834	2788	2743	2698	2653	2609	2565	2522	2479
44	2880	2834	2788	2742	2697	2652	2608	2564	2521	2478
45	2880	2833	2787	2741	2696	2652	2607	2564	2520	2477
46	2879	2832	2786	2741	2695	2651	2607	2563	2520	2477
47	2878	2831	2785	2740	2695	2650	2606	2562	2519	2476
48	2877	2831	2785	2739	2694	2649	2605	2561	2518	2475
49	2876	2830	2784	2738	2693	2649	2604	2561	2517	2474
50	2876	2829	2783	2737	2692	2648	2604	2560	2517	2474
51	2875	2828	2782	2737	2692	2647	2603	2559	2516	2473
52	2874	2828	2782	2736	2691	2646	2602	2558	2515	2472
53	2873	2827	2781	2735	2690	2646	2601	2558	2514	2472
54	2873	2826	2780	2735	2689	2645	2601	2557	2514	2471
55	2872	2825	2779	2734	2689	2644	2600	2556	2513	2470
56	2871	2824	2778	2733	2688	2643	2599	2556	2512	2470
57	2870	2824	2778	2732	2687	2643	2599	2555	2512	2469
58	2869	2823	2777	2731	2686	2642	2598	2554	2511	2468
59	2869	2822	2776	2731	2686	2641	2597	2553	2510	2467
60	2868	2821	2775	2730	2685	2640	2596	2553	2510	2467

"	h. ' 1.42	h. ' 1.43	h. ' 1.44	h. ' 1.45	h. ' 1.46	h. ' 1.47	h. ' 1.48	h. ' 1.49	h. ' 1.50	h. ' 1.51
0	2467	2424	2382	2341	2300	2259	2218	2178	2139	2099
1	2466	2424	2382	2340	2299	2258	2218	2178	2138	2099
2	2465	2423	2381	2339	2298	2257	2217	2177	2137	2098
3	2465	2422	2380	2339	2298	2257	2216	2176	2137	2098
4	2464	2421	2380	2338	2297	2256	2216	2176	2136	2097
5	2463	2421	2379	2337	2296	2255	2215	2175	2135	2096
6	2462	2420	2378	2337	2296	2255	2214	2174	2135	2096
7	2462	2419	2378	2336	2295	2254	2214	2174	2134	2095
8	2461	2419	2377	2335	2294	2253	2213	2173	2133	2094
9	2460	2418	2376	2335	2294	2253	2212	2172	2133	2094
10	2460	2417	2375	2334	2293	2252	2212	2172	2132	2093
11	2459	2417	2375	2333	2292	2251	2211	2171	2132	2092
12	2458	2416	2374	2333	2291	2251	2210	2170	2131	2092
13	2457	2415	2373	2332	2291	2250	2210	2170	2130	2091
14	2457	2414	2373	2331	2290	2249	2209	2169	2130	2090
15	2456	2414	2372	2331	2289	2249	2208	2169	2129	2090
16	2455	2413	2371	2330	2289	2248	2208	2168	2128	2089
17	2455	2412	2371	2329	2288	2247	2207	2167	2128	2088
18	2454	2412	2370	2328	2287	2447	2206	2167	2127	2088
19	2453	2411	2369	2328	2287	2246	2206	2166	2126	2087
20	2452	2410	2368	2327	2286	2245	2205	2165	2126	2086
21	2452	2410	2368	2326	2285	2245	2204	2165	2125	2086
22	2451	2409	2367	2326	2285	2244	2204	2164	2124	2085
23	2450	2408	2366	2325	2284	2243	2203	2163	2124	2084
24	2450	2408	2366	2324	2283	2243	2202	2163	2123	2084
25	2449	2407	2365	2324	2283	2242	2202	2162	2122	2083
26	2448	2406	2364	2323	2282	2241	2201	2161	2122	2083
27	2448	2405	2364	2322	2281	2241	2200	2161	2121	2082
28	2447	2405	2363	2322	2281	2240	2200	2160	2120	2081
29	2446	2404	2362	2321	2280	2239	2199	2159	2120	2081
30	2445	2403	2362	2320	2279	2239	2198	2159	2119	2081

LOGARITHMS.

"	h /	h /	h /	h /	h /	h /	h /	h /	h /	h /
	1.42	1.43	1.44	1.45	1.46	1.47	1.48	1.49	1.50	1.51
31	2445	2403	2361	2319	2279	2238	2198	2158	2118	2079
32	2444	2402	2360	2319	2278	2237	2197	2157	1118	2079
33	2443	2401	2359	2318	2277	2237	2196	2157	2117	2078
34	2443	2400	2359	2317	2276	2236	2196	2156	2115	2077
35	2442	2400	2358	2317	2276	2235	2195	2155	2115	2077
36	2441	2399	2357	2316	2275	2235	2194	2155	2115	2076
37	2440	2398	2357	2315	2274	2234	2194	2154	2114	2075
38	2440	2398	2356	2315	2274	2233	2193	2153	2114	2075
39	2439	2397	2355	2314	2273	2233	2192	2153	2113	2074
40	2438	2396	2355	2313	2272	2232	2192	2152	2113	2073
41	2438	2396	2354	2313	2272	2231	2191	2151	2112	2073
42	2437	2395	2353	2312	2271	2231	2190	2151	2111	2072
43	2436	2394	2353	2311	2270	2230	2190	2150	2111	2071
44	2436	2394	2352	2311	2270	2229	2189	2149	2110	2071
45	2435	2393	2351	2310	2269	2229	2188	2149	2109	2070
46	2434	2392	2350	2309	2268	2228	2188	2148	2109	2070
47	2433	2391	2350	2308	2268	2227	2187	2147	2108	2069
48	2433	2391	2349	2308	2267	2227	2186	2147	2107	2068
49	2432	2390	2348	2307	2266	2226	2186	2146	2107	2069
50	2431	2389	2348	2306	2266	2225	2185	2145	2106	2067
51	2431	2389	2347	2306	2265	2225	2184	2145	2105	2066
52	2430	2388	2346	2305	2264	2224	2184	2144	2105	2066
53	2429	2387	2346	2304	2264	2223	2183	2143	2104	2065
54	2429	2387	2345	2304	2263	2223	2182	2143	2103	2064
55	2428	2386	2344	2303	2262	2222	2182	2142	2103	2064
56	2427	2385	2344	2302	2262	2221	2181	2141	2102	2063
57	2426	2384	2343	2302	2261	2220	2180	2141	2101	2062
58	2426	2384	2342	2301	2260	2220	2180	2140	2101	2062
59	2425	2383	2341	2300	2260	2219	2179	2139	2100	2061
60	2424	2382	2341	2300	2259	2218	2178	2139	2099	2061

E

PROPORTIONAL

''	h ' 1.52	h ' 1.53	h ' 1.54	h ' 1.55	h ' 1.56	h ' 1.57	h ' 1.58	h ' 1.59	h ' 2.0	h ' 2.1
c	2061	2022	1984	1946	1908	1871	1834	1797	1761	1725
1	2060	2021	1983	1945	1907	1870	1833	1797	1760	1724
2	2059	2021	1982	1944	1907	1870	1833	1796	1760	1724
3	2059	2020	1982	1944	1906	1869	1832	1795	1759	1723
4	3058	2019	1981	1943	1906	1868	1831	1795	1758	1722
5	2057	2019	1980	1943	1905	1868	1831	1794	1758	1722
6	2057	2018	1980	1942	1904	1867	1830	1794	1757	1721
7	2056	2017	1979	1941	1904	1867	1830	1793	1757	1721
8	2055	2017	1979	1941	1903	1866	1829	1792	1756	1720
9	2055	2016	1978	1940	1903	1865	1828	1792	1755	1719
10	2054	2016	1977	1939	1902	1865	1828	1791	1755	1719
11	2053	2015	1977	1939	1901	1864	1827	1791	1754	1718
12	2053	2014	1976	1938	1901	1863	1827	1790	1754	1718
13	2052	2014	1975	1938	1900	1863	1826	1789	1753	1717
14	2051	2013	1975	1937	1899	1862	1825	1789	1752	1716
15	2051	2012	1974	1936	1899	1862	1825	1788	1752	1716
16	2050	2012	1973	1936	1898	1861	1824	1787	1751	1715
17	2050	2011	1973	1935	1898	1860	1823	1787	1751	1715
18	2049	2010	1972	1934	1897	1860	1823	1786	1750	1714
19	2048	2010	1972	1934	1896	1859	1822	1786	1749	1713
20	2048	2009	1971	1933	1896	1858	1822	1785	1749	1713
21	3047	2009	1970	1933	1895	1858	1821	1785	1748	1712
22	2046	2008	1970	1932	1894	1857	1820	1784	1748	1712
23	2046	2007	1969	1931	1894	1857	1820	1783	1747	1711
24	2045	2007	1968	1931	1893	1856	1819	1783	1746	1711
25	2044	2006	1968	1930	1893	1855	1819	1782	1746	1710
26	2044	2005	1967	1929	1892	1855	1818	1781	1745	1709
27	2043	2005	1967	1929	1891	1854	1817	1781	1745	1709
28	2042	2004	1966	1928	1891	1854	1817	1780	1744	1708
29	2042	2004	1965	1927	1890	1853	1816	1780	1743	1708
30	2041	2003	1965	1927	1889	1852	1816	1779	1743	1707

LOGARITHMS.

	h o 1.52	h o 1.53	h o 1.54	h o 1.55	h o 1.56	h o 1.57	h o 1.58	h o 1.59	h o 2. 0	h o 2. 1
31	2041	2002	1964	1926	1880	1852	1815	1778	1742	1706
32	2040	2001	1963	1926	1888	1851	1814	1778	1742	1706
33	2039	2001	1963	1925	1888	1850	1814	1777	1741	1705
34	2039	2000	1962	1924	1887	1850	1813	1777	1740	1705
35	2038	2000	1961	1924	1886	1849	1812	1776	1740	1704
36	2037	1999	1961	1923	1886	1849	1812	1775	1739	1703
37	2037	1998	1960	1922	1885	1848	1811	1775	1739	1703
38	2036	1998	1960	1922	1884	1847	1811	1774	1738	1702
39	2035	1997	1959	1921	1884	1847	1810	1774	1737	1702
40	2035	1996	1958	1921	1883	1846	1809	1773	1737	1701
41	2034	1996	1958	1920	1883	1846	1809	1772	1736	1700
42	2033	1995	1957	1919	1882	1845	1808	1772	1736	1700
43	2033	1994	1956	1919	1881	1844	1808	1771	1735	1699
44	2032	1994	1956	1918	1881	1844	1807	1771	1734	1699
45	2032	1993	1955	1918	1880	1843	1806	1770	1734	1698
46	2031	1993	1955	1917	1879	1842	1806	1769	1733	1697
47	2030	1992	1954	1916	1879	1842	1805	1769	1733	1697
48	2030	1991	1953	1916	1878	1841	1805	1768	1732	1696
49	2029	1991	1953	1915	1878	1841	1804	1768	1731	1696
50	2028	1990	1952	1914	1877	1840	1803	1767	1731	1695
51	2028	1989	1951	1914	1876	1839	1803	1766	1730	1694
52	2027	1989	1951	1913	1876	1839	1802	1766	1730	1694
53	2026	1988	1950	1912	1875	1838	1801	1765	1729	1693
54	2026	1987	1950	1912	1875	1838	1801	1765	1728	1693
55	2025	1987	1949	1911	1874	1837	1800	1764	1728	1692
56	2024	1986	1948	1911	1873	1836	1800	1763	1727	1691
57	2024	1986	1948	1910	1873	1836	1799	1763	1727	1691
58	2023	1985	1947	1909	1872	1835	1798	1762	1726	1690
59	2023	1984	1946	1909	1871	1834	1798	1761	1725	1690
60	2022	1984	1946	1908	1871	1834	1797	1761	1725	1689

PROPORTIONAL

"	h ' 2. 2	h ' 2. 3	h ' 2. 4	h ' 2. 5	h ' 2. 6	h ' 2. 7	h ' 2. 8	h ' 2. 9	h ' 2. 10	h ' 2. 11
0	1689	1654	1619	1584	1549	1515	1481	1447	1413	1380
1	1688	1653	1618	1583	1542	1514	1480	1446	1413	1379
2	1688	1652	1617	1582	1548	1514	1479	1446	1412	1379
3	1687	1652	1617	1582	1547	1513	1479	1445	1412	1378
4	1687	1651	1616	1581	1547	1512	1478	1445	1411	1378
5	1686	1651	1616	1581	1546	1512	1478	1444	1410	1377
6	1686	1650	1615	1580	1546	1511	1477	1443	1410	1377
7	1685	1650	1614	1580	1545	1511	1477	1443	1409	1376
8	1684	1649	1614	1579	1544	1510	1476	1442	1409	1376
9	1684	1648	1613	1578	1544	1510	1476	1442	1408	1375
10	1683	1648	1613	1578	1543	1509	1475	1441	1408	1374
11	1683	1647	1612	1577	1543	1508	1474	1441	1407	1374
12	1682	1647	1612	1577	1542	1508	1474	1440	1407	1373
13	1681	1646	1611	1576	1542	1507	1473	1440	1406	1373
14	1681	1645	1610	1575	1541	1507	1473	1439	1405	1372
15	1680	1645	1610	1575	1540	1506	1472	1438	1405	1372
16	1680	1644	1609	1574	1540	1506	1472	1438	1404	1371
17	1679	1644	1609	1574	1539	1505	1471	1437	1404	1371
18	1678	1643	1608	1573	1539	1504	1470	1437	1403	1370
19	1678	1642	1607	1573	1538	1504	1470	1436	1403	1369
20	1677	1642	1607	1572	1538	1503	1469	1436	1402	1369
21	1677	1641	1606	1571	1537	1503	1469	1435	1402	1368
22	1676	1641	1606	1571	1536	1502	1468	1434	1401	1368
23	1675	1640	1605	1570	1536	1502	1468	1434	1400	1367
24	1675	1640	1605	1570	1535	1501	1467	1433	1400	1367
25	1674	1639	1604	1569	1535	1500	1466	1433	1399	1366
26	1674	1638	1603	1569	1534	1500	1466	1432	1399	1366
27	1673	1638	1603	1568	1534	1499	1465	1432	1398	1365
28	1673	1637	1602	1567	1533	1499	1465	1431	1398	1365
29	1672	1637	1602	1567	1532	1498	1464	1431	1397	1364
30	1671	1636	1601	1566	1532	1498	1464	1430	1397	1363

LOGARITHMS.

"	h ' 2. 2	h ' 2. 3	h ' 2. 4	h ' 2. 5	h ' 2. 6	h ' 2. 7	h ' 2. 8	h ' 2. 9	h ' 2. 10	h ' 2. 11
31	1671	1635	1600	1566	1531	1497	1463	1429	1396	1363
32	1670	1635	1600	1565	1531	1496	1463	1429	1395	1362
33	1670	1634	1599	1565	1530	1496	1462	1428	1395	1362
34	1669	1634	1599	1564	1529	1495	1461	1428	1394	1361
35	1668	1633	1598	1563	1529	1495	1461	1427	1394	1361
36	1668	1633	1598	1563	1528	1494	1460	1427	1393	1360
37	1667	1632	1597	1562	1528	1494	1460	1426	1393	1360
38	1667	1631	1596	1562	1527	1493	1459	1426	1392	1359
39	1666	1631	1596	1561	1527	1493	1459	1425	1392	1359
40	1665	1630	1595	1560	1526	1492	1458	1424	1391	1358
41	1665	1630	1595	1560	1525	1491	1457	1424	1390	1357
42	1664	1629	1594	1559	1525	1491	1457	1423	1390	1357
43	1664	1628	1593	1559	1524	1490	1456	1423	1389	1356
44	1663	1628	1593	1558	1524	1490	1456	1422	1389	1356
45	1663	1627	1592	1558	1523	1489	1455	1422	1388	1355
46	1662	1627	1592	1557	1523	1489	1455	1421	1388	1355
47	1661	1626	1591	1556	1522	1488	1454	1420	1387	1354
48	1661	1626	1591	1556	1522	1487	1454	1420	1387	1354
49	1660	1625	1590	1555	1521	1487	1453	1419	1386	1353
50	1660	1624	1589	1555	1520	1486	1452	1419	1386	1352
51	1659	1624	1589	1554	1520	1486	1452	1418	1385	1352
52	1658	1623	1588	1554	1519	1485	1451	1418	1384	1351
53	1658	1623	1588	1553	1518	1485	1451	1417	1384	1351
54	1657	1622	1587	1552	1518	1484	1450	1417	1383	1350
55	1657	1621	1586	1552	1518	1483	1450	1416	1383	1350
56	1656	1621	1586	1551	1517	1483	1449	1415	1382	1349
57	1655	1620	1585	1551	1516	1482	1449	1415	1382	1349
58	1655	1620	1585	1550	1516	1482	1448	1414	1381	1348
59	1654	1619	1584	1550	1515	1481	1447	1414	1381	1347
60	1654	1619	1584	1549	1515	1481	1447	1413	1380	1347

"	h₀ ' 2.12	h₀ ' 2.13	h₀ ' 2.14	h₀ ' 2.15	h₀ ' 2.16	h₀ ' 2.17	h₀ ' 2.18	h₀ ' 2.19	h₀ ' 2.20	h₀ ' 2.21
0	1347	1314	1282	1249	1217	1186	1154	1123	1091	1061
1	1346	1314	1281	1249	1217	1185	1153	1122	1091	1060
2	1346	1313	1281	1248	1216	1184	1153	1121	1090	1059
3	1345	1313	1280	1248	1216	1184	1152	1121	1090	1059
4	1345	1312	1279	1247	1215	1183	1152	1120	1089	1058
5	1344	1311	1279	1247	1215	1183	1151	1120	1089	1058
6	1344	1311	1278	1246	1214	1182	1151	1119	1088	1057
7	1343	1310	1278	1246	1214	1182	1150	1119	1088	1057
8	1343	1310	1277	1245	1213	1181	1150	1118	1087	1056
9	1342	1309	1277	1245	1213	1181	1149	1118	1087	1056
10	1341	1309	1276	1244	1212	1180	1149	1117	1086	1055
11	1341	1308	1276	1243	1211	1180	1148	1117	1086	1055
12	1340	1308	1275	1243	1211	1179	1148	1116	1085	1054
13	1340	1307	1275	1242	1210	1179	1147	1116	1085	1054
14	1339	1307	1274	1242	1210	1178	1147	1115	1084	1053
15	1339	1306	1274	1241	1209	1178	1146	1115	1084	1053
16	1338	1305	1273	1241	1209	1177	1146	1114	1083	1052
17	1338	1305	1272	1240	1208	1177	1145	1114	1083	1052
18	1337	1304	1272	1240	1208	1176	1145	1113	1062	1051
19	1337	1304	1271	1239	1207	1175	1144	1113	1082	1051
20	1336	1303	1271	1239	1207	1175	1143	1112	1081	1050
21	1335	1303	1270	1238	1206	1174	1143	1112	1081	1050
22	1335	1302	1270	1238	1206	1174	1142	1111	1080	1049
23	1334	1302	1269	1237	1205	1173	1142	1111	1080	1049
24	1334	1301	1269	1237	1205	1173	1141	1110	1079	1048
25	1333	1301	1268	1236	1204	1172	1141	1110	1079	1048
26	1333	1300	1268	1235	1203	1172	1140	1109	1078	1047
27	1332	1300	1267	1235	1203	1171	1140	1109	1078	1047
28	1332	1299	1267	1234	1202	1171	1139	1108	1077	1046
29	1331	1298	1266	1234	1202	1170	1139	1107	1076	1046
30	1331	1298	1266	1233	1201	1170	1138	1107	1076	1045

LOGARITHMS.

"	2. 12	2. 13	2. 14	2. 15	2. 16	2. 17	2. 18	2. 19	2. 20	2. 21
31	1330	1297	1265	1233	1201	1169	1138	1106	1075	1045
32	1329	1297	1264	1232	1200	1169	1137	1106	1075	1044
33	1329	1296	1264	1232	1200	1168	1137	1105	1074	1044
34	1328	1296	1263	1231	1199	1168	1136	1105	1074	1043
35	1328	1295	1263	1231	1199	1167	1136	1104	1073	1043
36	1327	1295	1262	1230	1198	1167	1135	1104	1073	1042
37	1327	1294	1262	1230	1198	1166	1135	1103	1072	1042
38	1326	1294	1261	1229	1197	1165	1134	1103	1072	1041
39	1326	1293	1261	1229	1197	1165	1134	1102	1071	1041
40	1325	1292	1260	1228	1196	1164	1133	1102	1071	1040
41	1325	1292	1260	1227	1196	1164	1132	1101	1070	1039
42	1324	1291	1259	1227	1195	1163	1132	1101	1070	1039
43	1323	1291	1258	1226	1194	1163	1131	1100	1069	1038
44	1323	1290	1258	1226	1194	1162	1131	1100	1069	1038
45	1322	1290	1257	1225	1193	1162	1130	1099	1068	1037
46	1322	1289	1257	1225	1193	1161	1130	1099	1068	1037
47	1321	1289	1256	1224	1192	1161	1129	1098	1067	1036
48	1321	1288	1256	1224	1192	1160	1129	1098	1067	1036
49	1320	1288	1255	1223	1191	1160	1128	1097	1066	1035
50	1320	1287	1255	1223	1191	1159	1128	1097	1066	1035
51	1319	1287	1254	1222	1190	1159	1127	1096	1065	1034
52	1319	1286	1254	1222	1190	1158	1127	1096	1065	1034
53	1318	1285	1253	1221	1189	1158	1126	1095	1064	1033
54	1317	1285	1253	1221	1189	1157	1126	1095	1064	1033
55	1317	1284	1252	1220	1188	1157	1125	1094	1063	1032
56	1316	1284	1251	1219	1188	1156	1125	1093	1063	1032
57	1316	1283	1251	1219	1187	1156	1124	1093	1062	1031
58	1315	1283	1250	1218	1187	1155	1124	1092	1062	1031
59	1315	1282	1250	1218	1186	1154	1123	1092	1061	1030
60	1314	1282	1249	1217	1186	1154	1123	1091	1061	1030

"	2.22	2.23	2.24	2.25	2.26	2.27	2.28	2.29	2.30	2.31
0	1030	0999	0969	0939	0909	0880	0850	0821	0792	0763
1	1029	0999	0969	0939	0909	0879	0850	0820	0791	0762
2	1029	0998	0968	0938	0908	0879	0849	0820	0791	0762
3	1028	0998	0968	0938	0908	0878	0849	0819	0790	0762
4	1028	0997	0967	0937	0907	0878	0848	0819	0790	0761
5	1027	0997	0967	0937	0907	0877	0848	0818	0789	0761
6	1027	0996	0966	0936	0906	0877	0847	0818	0789	0760
7	1026	0996	0966	0936	0906	0876	0847	0817	0788	0760
8	1026	0995	0965	0935	0905	0876	0846	0817	0788	0759
9	1025	0995	0965	0935	0905	0875	0846	0816	0787	0759
10	1025	0994	0964	0934	0904	0875	0845	0816	0787	0758
11	1024	0994	0964	0934	0904	0874	0845	0815	0787	0758
12	1024	0993	0963	0933	0903	0874	0844	0815	0786	0757
13	1023	0993	0963	0933	0903	0873	0844	0814	0786	0757
14	1023	0992	0962	0932	0902	0873	0843	0814	0785	0756
15	1022	0992	0962	0932	0902	0872	0843	0814	0785	0756
16	1022	0991	0961	0931	0901	0872	0842	0813	0784	0755
17	1021	0991	0961	0931	0901	0871	0842	0813	0784	0755
18	1021	0990	0960	0930	0900	0871	0841	0812	0783	0754
19	1020	0990	0960	0930	0900	0870	0841	0812	0783	0754
20	1020	0989	0959	0929	0899	0870	0840	0811	0782	0753
21	1019	0989	0959	0929	0899	0869	0840	0811	0782	0753
22	1019	0988	0958	0928	0898	0869	0839	0810	0781	0752
23	1018	0988	0958	0928	0898	0868	0839	0810	0781	0752
24	1018	0987	0957	0927	0897	0868	0838	0809	0780	0751
25	1017	0987	0957	0927	0897	0867	0838	0809	0780	0751
26	1017	0986	0956	0926	0896	0867	0837	0808	0779	0750
27	1016	0986	0956	0926	0896	0866	0837	0808	0779	0750
28	1016	0985	0955	0925	0895	0866	0836	0807	0778	0750
29	1015	0985	0955	0925	0895	0865	0836	0807	0778	0749
30	1015	0984	0954	0924	0894	0865	0835	0806	0777	0749

	h.	h.	h.	h.	h.	h.	h.	h.	h.	h.
''	2.22	2.23	2.24	2.25	2.26	2.27	2.28	2.29	2.30	2.31
31	1014	0984	0954	0924	0894	0864	0835	0806	0777	0748
32	1014	0983	0953	0923	0893	0864	0834	0805	0776	0748
33	1013	0983	0953	0923	0893	0863	0834	0805	0776	0747
34	1013	0982	0952	0922	0892	0863	0833	0804	0775	0747
35	1012	0982	0952	0922	0892	0862	0833	0804	0775	0746
36	1012	0981	0951	0921	0891	0862	0833	0803	0774	0746
37	1011	0981	0951	0921	0891	0861	0832	0803	0774	0745
38	1010	0980	0950	0920	0890	0861	0832	0802	0773	0745
39	1010	0980	0950	0920	0890	0860	0831	0802	0773	0744
40	1009	0979	0949	0919	0889	0860	0831	0801	0773	0744
41	1009	0979	0949	0919	0889	0859	0830	0801	0772	0743
42	1008	0978	0948	0918	0888	0859	0830	0801	0772	0743
43	1008	0978	0948	0918	0888	0858	0829	0800	0771	0742
44	1007	0977	0947	0917	0887	0858	0829	0800	0771	0742
45	1007	0977	0947	0917	0887	0857	0828	0799	0770	0741
46	1006	0976	0946	0916	0886	0857	0828	0799	0770	0741
47	1006	0976	0946	0916	0886	0856	0827	0798	0769	0740
48	1005	0975	0945	0915	0885	0856	0827	0798	0769	0740
49	1005	0975	0945	0915	0885	0855	0826	0797	0768	0739
50	1004	0974	0944	0914	0884	0855	0826	0797	0768	0739
51	1004	0974	0944	0914	0884	0855	0825	0796	0767	0739
52	1003	0973	0943	0913	0883	0854	0825	0796	0767	0738
53	1003	0973	0943	0913	0883	0854	0824	0795	0766	0738
54	1002	0972	0942	0912	0883	0853	0824	0795	0766	0737
55	1002	0972	0942	0912	0882	0853	0823	0794	0765	0737
56	1001	0971	0941	0911	0882	0852	0823	0794	0765	0736
57	1001	0971	0941	0911	0881	0852	0822	0793	0764	0736
58	1000	0970	0940	0910	0881	0851	0822	0793	0764	0735
59	1000	0970	0940	0910	0880	0851	0821	0792	0763	0735
60	0999	0969	0939	0909	0880	0850	0821	0792	0763	0734

F

PROPORTIONAL

	h. '	h. '	h. '	h. '	h. '	h. '	h. '	h. '	h. '	h. '
"	2. 32	2. 33	2. 34	2· 35	2. 36	2. 37	2. 38	2. 39	2. 40	2. 41
0	0734	0706	0678	0649	0621	0594	0566	0539	0512	0484
1	0734	0705	0677	0649	0621	0593	0566	0538	0511	0484
2	0733	0705	0677	0648	0621	0592	0565	0538	0511	0484
3	0733	0704	0676	0648	0620	0592	0565	0537	0510	0483
4	0732	0704	0676	0648	0620	0592	0564	0537	0510	0483
5	0732	0703	0675	0647	0619	0591	0564	0536	0509	0482
6	0731	0703	0675	0647	0619	0591	0563	0536	0509	0482
7	0731	0702	0674	0646	0618	0590	0563	0536	0508	0481
8	0730	0702	0674	0646	0618	0590	0562	0535	0508	0481
9	0730	0702	0673	0645	0617	0590	0562	0535	0507	0480
10	0729	0701	0673	0645	0617	0589	0562	0534	0507	0480
11	0729	0701	0672	0644	0616	0589	0561	0534	0507	0479
12	0729	0700	0672	0644	0616	0588	0561	0533	0506	0479
13	0728	0700	0671	0643	0615	0588	0560	0533	0506	0479
14	0728	0699	0671	0643	0615	0587	0560	0532	0505	0478
15	0727	0699	0670	0642	0615	0587	0559	0532	0505	0478
16	0727	0698	0670	0642	0614	0586	0559	0531	0504	0477
17	0726	0698	0669	0641	0614	0986	0558	0531	0504	0477
18	0726	0697	0669	0641	0613	0585	0558	0531	0503	0476
19	0725	0697	0669	0641	0613	0585	0557	0530	0503	0476
20	0725	0696	0668	0640	0612	0584	0557	0530	0502	0475
21	0724	0696	0668	0640	0612	0584	0557	0529	0502	0475
22	0724	0695	0667	0639	0611	0584	0556	0529	0501	0475
23	0723	0695	0667	0639	0611	0583	0556	0528	0501	0474
24	0723	0694	0666	0638	0610	0583	0555	0528	0501	0474
25	0722	0694	0666	0638	0610	0584	0555	0527	0500	0473
26	0722	0693	0665	0637	0609	0582	0554	0527	0500	0473
27	0721	0693	0665	0637	0609	0581	0554	0526	0499	0472
28	0721	0693	0664	0636	0608	0581	0553	0526	0499	0472
29	0720	0692	0664	0636	0608	0581	0553	0526	0498	0471
30	0720	0692	0663	0635	0608	0580	0552	0525	0498	0471

"	h. ' 2.32	h. ' 2.33	h. ' 2.34	h. ' 2.35	h. ' 2.36	h. ' 2.37	h. ' 2.38	h. ' 2.39	h. ' 2.40	h. ' 2.41
31	0720	0691	0663	0635	0607	0579	0552	0525	0497	0471
32	0719	0691	0662	0634	0607	0579	0551	0524	0497	0470
33	0719	0690	0662	0634	0606	0579	0551	0524	0497	0470
34	0718	0690	0662	0634	0606	0578	0551	0523	0496	0469
35	0718	0689	0661	0633	0605	0578	0550	0523	0496	0469
36	0717	0689	0661	0633	0605	0577	0550	0522	0495	0468
37	0717	0688	0660	0632	0604	0577	0549	0522	0495	0468
38	0716	0688	0660	0632	0604	0576	0549	0521	0494	0467
39	0716	0687	0659	0631	0603	0576	0548	0521	0494	0467
40	0715	0687	0659	0631	0603	0575	0548	0521	0493	0466
41	0715	0686	0658	0630	0602	0575	0547	0520	0493	0466
42	0714	0686	0658	0630	0602	0574	0547	0520	0493	0466
43	0714	0685	0657	0629	0602	0574	0546	0519	0492	0465
44	0713	0685	0657	0629	0601	0573	0546	0519	0492	0465
45	0713	0685	0656	0628	0601	0573	0546	0518	0491	0464
46	0712	0684	0656	0628	0600	0573	0545	0518	0491	0464
47	0712	0684	0655	0627	0600	0572	0545	0517	0490	0463
48	0711	0683	0655	0627	0599	0572	0544	0517	0490	0463
49	0711	0683	0655	0627	0599	0571	0544	0516	0489	0462
50	0711	0682	0654	0626	0598	0571	0543	0516	0489	0462
51	0710	0682	0654	0626	0598	0570	0543	0516	0489	0462
52	0710	0681	0653	0625	0597	0570	0542	0515	0488	0461
53	0709	0681	0653	0625	0597	0569	0542	0515	0488	0461
54	0709	0680	0652	0624	0596	0569	0541	0514	0487	0460
55	0708	0680	0652	0624	0596	0568	0541	0514	0487	0460
56	0708	0679	0651	0623	0596	0568	0541	0513	0486	0459
57	0707	0679	0651	0623	0595	0568	0540	0513	0486	0459
58	0707	0678	0650	0622	0595	0567	0540	0512	0485	0458
59	0706	0678	0650	0622	0594	0567	0539	0512	0485	0458
60	0706	0678	0649	0621	0594	0566	0539	0512	0484	0458

h; '	h. '	h. '	h. '	h. '	h. '	h. '	h. '	h	h. '
" 2. 42	2. 43	2. 44	2. 45	2. 46	2. 47	2. 48	2. 49	2. 50	2. 51
0 0458	0431	0404	0378	0352	0326	0300	0274	0248	0223
1 0457	0430	0404	0377	0351	0325	0299	0273	0248	0222
2 0457	0430	0403	0377	0351	0325	0299	0273	0247	0222
3 0456	0430	0403	0377	0350	0324	0298	0273	0247	0221
4 0456	0429	0402	0376	0350	0324	0298	0272	0246	0221
5 0455	0429	0402	0376	0349	0323	0297	0272	0246	0221
6 0455	0428	0402	0375	0349	0323	0297	0271	0246	0220
7 0454	0428	0401	0375	0349	0322	0297	0271	0245	0220
8 0454	0427	0401	0374	0348	0322	0296	0270	0245	0219
9 0454	0427	0400	0374	0348	0322	0296	0270	0244	0219
10 0453	0426	0400	0373	0347	0321	0295	0270	0244	0218
11 0453	0426	0399	0373	0347	0321	0295	0269	0244	0218
12 0452	0426	0399	0373	0346	0320	0294	0269	0243	0216
13 0452	0425	0399	0372	0346	0320	0294	0268	0243	0217
14 0451	0425	0398	0372	0346	0319	0294	0268	0242	0217
15 0451	0424	0398	0371	0345	0319	0293	0267	0242	0216
16 0450	0424	0397	0371	0345	0319	0293	0267	0241	0216
17 0450	0423	0397	0370	0344	0318	0292	0267	0241	0216
18 0450	0423	0396	0370	0344	0318	0292	0266	0241	0215
19 0449	0423	0396	0370	0343	0317	0291	0266	0240	0215
20 0449	0422	0395	0369	0343	0317	0291	0265	0240	0214
21 0448	0422	0395	0369	0342	0316	0291	0265	0239	0214
22 0448	0421	0395	0368	0342	0316	0290	0264	0159	021.
23 0447	0421	0394	0368	0342	0316	0290	0264	0238	0213
24 0447	0420	0394	0367	0341	0315	0289	0264	0238	0213
25 0446	0420	0393	0367	0341	0315	0289	0263	0238	0212
26 0446	0419	0393	0366	0340	0314	0288	0263	0237	0212
27 0445	0411	0392	0366	0340	0314	0288	0262	0237	0211
28 0445	0411	0392	0366	0339	0313	0288	0262	0236	0211
29 0445	0418	0391	0365	0339	0313	0287	0261	0236	0210
30 0444	0418	0391	0365	0339	0313	0287	0261	0235	0210

LOGARITHMS.

"	2.42	2.43	2.44	2.45	2.46	2.47	2.48	2.49	2.50	2.51
31	0444	0417	0391	0364	0338	0312	0286	0261	0235	0210
32	0443	0417	0390	0364	0338	0312	0286	0260	0235	0209
33	0443	0416	0390	0363	0337	0311	0285	0260	0234	0209
34	0442	0416	0389	0363	0337	0311	0285	0259	0234	0208
35	0442	0415	0389	0363	0336	0310	0285	0259	0233	0208
36	0442	0415	0388	0362	0336	0310	0284	0258	0233	0208
37	0441	0414	0388	0362	0336	0310	0284	0258	0232	0207
38	0441	0414	0388	0361	0335	0309	0283	0258	0232	0207
39	0440	0414	0387	0361	0335	0309	0283	0257	0232	0206
40	0440	0413	0387	0360	0334	0308	0282	0257	0231	0206
41	0439	0413	0386	0360	0334	0308	0282	0256	0231	0205
42	0439	0412	0386	0359	0333	0307	0282	0256	0230	0205
43	0438	0412	0385	0359	0333	0307	0281	0255	0230	0205
44	0438	0411	0385	0359	0332	0306	0281	0255	0230	0204
45	0438	0411	0384	0358	0332	0306	0280	0255	0229	0204
46	0437	0410	0384	0358	0332	0306	0280	0254	0229	0203
47	0437	0410	0384	0357	0331	0305	0279	0254	0228	0203
48	0436	0410	0383	0357	0331	0305	0279	0253	0228	0202
49	0436	0409	0383	0356	0330	0304	0279	0253	0227	0202
50	0435	0409	0382	0356	0330	0304	0278	0252	0227	0202
51	0435	0406	0381	0356	0329	0304	0278	0252	0227	0201
52	0434	0406	0381	0355	0329	0303	0277	0252	0226	0201
53	0434	0407	0381	0355	0329	0303	0277	0251	0226	0200
54	0434	0407	0381	0354	0328	0302	0276	0251	0225	0200
55	0433	0406	0380	0354	0328	0302	0276	0250	0225	0200
56	0433	0406	0380	0353	0327	0301	0276	0250	0224	0199
57	0432	0406	0379	0353	0327	0301	0275	0250	0224	0199
58	0432	0405	0379	0352	0326	0300	0275	0249	0224	0198
59	0431	0405	0378	0352	0326	0300	0274	0249	0225	0198
60	0431	0404	0378	0352	0326	0300	0274	0248	0223	0197

h. /	h. /	h. /	h. /	h. /	h. /	h. /	h. /
" 2. 52	2. 53	2. 54	2. 55	2. 56	2. 57	2. 58	2. 59
0 0197	0172	0147	0122	0098	0073	0049	0024
1 0197	0172	0147	0122	0097	0073	0048	0024
2 0197	0171	0146	0121	0097	0072	0048	0023
3 0196	0171	0146	0121	0096	0072	0047	0023
4 0196	0171	0146	0121	0096	0071	0047	0023
5 0195	0170	0145	0120	0096	0071	0046	0022
6 0195	0170	0145	0120	0095	0071	0046	0022
7 0194	0169	0144	0119	0095	0070	0046	0021
8 0194	0169	0144	0119	0094	0070	0045	0021
9 0194	0169	0143	0119	0094	0069	0045	0021
10 0193	0168	0143	0118	0093	0069	0044	0020
11 0193	0168	0143	0118	0093	0068	0044	0020
12 0192	0167	0142	0117	0093	0068	0044	0019
13 0192	0167	0142	0117	0092	0068	0043	0019
14 0192	0166	0141	0117	0092	0067	0043	0018
15 0191	0166	0141	0116	0091	0067	0042	0018
16 0191	0166	0141	0116	0091	0066	0042	0018
17 0190	0165	0140	0115	0091	0066	0042	0017
18 0190	0165	0140	0115	0090	0066	0041	0017
19 0189	0164	0139	0114	0090	0065	0041	0016
20 0189	0164	0139	0114	0089	0065	0040	0016
21 0189	0163	0139	0114	0089	0064	0040	0016
22 0188	0163	0138	0113	0089	0064	0040	0015
23 0188	0163	0138	0113	0088	0064	0039	0015
24 0187	0162	0137	0112	0088	0063	0039	0015
25 0187	0162	0137	0112	0087	0063	0038	0014
26 0186	0161	0136	0112	0087	0062	0038	0014
27 0186	0161	0136	0111	0087	0062	0038	0013
28 0186	0161	0136	0111	0086	0062	0037	0013
29 0185	0160	0135	0110	0086	0061	0037	0012
30 0185	0160	0135	0110	0085	0061	0036	0012

LOGARITHMS.

"	h. ∘ ′ 2.52	h. ∘ ′ 2.53	h. ∘ ′ 2.54	h. ∘ ′ 2.55	h. ∘ ′ 2.56	h. ∘ ′ 2.57	h. ∘ ′ 2.58	h. ∘ ′ 2.59
31	0184	0159	0134	0110	0085	0060	0036	0012
32	0184	0159	0134	0109	0084	0060	0035	0011
33	0184	0158	0134	0109	0084	0060	0035	0011
34	0183	0158	0133	0108	0084	0059	0035	0010
35	0183	0158	0133	0108	0083	0059	0034	0010
36	0182	0157	0132	0107	0083	0058	0034	0010
37	0182	0157	0132	0107	0082	0058	0033	0009
38	0181	0156	0131	0107	0082	0057	0033	0009
39	0181	0156	0131	0106	0082	0057	0033	0008
40	0181	0156	0131	0106	0081	0057	0032	0008
41	0180	0155	0130	0105	0081	0056	0032	0008
42	0180	0155	0130	0105	0080	0056	0031	0007
43	0179	0154	0129	0105	0080	0055	0031	0007
44	0179	0154	0129	0104	0080	0055	0031	0006
45	0179	0153	0129	0104	0079	0055	0030	0006
46	0178	0153	0128	0103	0079	0054	0030	0006
47	0178	0153	0128	0103	0078	0054	0029	0005
48	0177	0152	0127	0103	0078	0053	0029	0005
49	0177	0152	0127	0102	0077	0053	0029	0004
50	0176	0151	0126	0102	0077	0053	0028	0004
51	0176	0151	0126	0101	0077	0052	0028	0004
52	0176	0151	0126	0101	0076	0052	0027	0003
53	0175	0150	0125	0100	0076	0051	0027	0003
54	0175	0150	0125	0100	0075	0051	0027	0002
55	0174	0149	0124	0100	0075	0051	0026	0002
56	0174	0149	0124	0099	0075	0050	0026	0002
57	0174	0148	0124	0099	0074	0050	0025	0001
58	0173	0148	0123	0098	0074	0049	0025	0001
59	0173	0148	0123	0098	0073	0049	0025	0000
60	0172	0147	0122	0098	0073	0049	0024	0000

EXPLICATION and USE

OF THE

T A B L E S

Requifite to be ufed with the Astronomical and Nautical Ephemeris.

THE Rays of Light in paffing through the Atmofphere being bent out of their ftrait Courfe into a
curved Line, it thence happens that all the heavenly Bodies, except when they are in the Zenith, appear higher
than they ought to do, and fo much the more, the nearer
they are to the Horizon. Hence they appear to rife fome Minutes fooner, and fet fome Minutes later than they would
do, if there was no Atmofphere, or if it had not this
Power of turning the Rays of Light out of their Courfe.
This apparent Elevation of the heavenly Bodies above their
true Height is called the Refraction of their Light, or, in
common Speaking, the Refraction of the Objects. The Effect of it is contained in Table Page 2d. and is fuited to a
mean Temperature of the Air at Greenwich; the Height
of the Barometer being 29,4 Inches, and that of the Thermometer of Fahrenheit's Conftruction 50 Degrees; or,
which comes to the fame Thing, 30 Inches of the Barometer, and 55 of the Thermometer. It is deduced from
a Rule invented by Dr. Bradley, and by him adapted to his
Obfervations, that the Refraction at any Altitude, is to
57", the Refraction at the Altitude of 45°; as the Tangent
of the apparent Zenith Diftance leffened by three times the

S   Refraction

Refraction taken out of any common Table, is to the Radius. To allow for the Variations of Refraction in different Temperatures of the Air, he has stated another Rule, derived also from or confirmed by his Observations, that the true Refraction is to that expressed by his first Rule, or contained in this Table, in a direct Ratio of the Altitude of the Barometer to 29⁴₅ Inches; and in an Inverse Ratio of the Altitude of the Thermometer increased by 350, to the Number 400.

It is evident that all observed Altitudes of the heavenly Bodies ought to be diminished by the Numbers taken out of this Table, particularly the Meridian Altitudes of the Sun and Stars, &c. the Altitudes of the Sun and Stars designed for computing the apparent Time of the Day, and the Altitudes of the Sun taken for computing his Azimuth. The Time for taking an Amplitude of the Sun is not when he appears in the Horizon, but when his Centre appears 29' high, or his lower Limb 15'; or upper Limb 43' above the true Horizon; but the Quantity of the Dip, p. 14. is to be added to these Numbers to find the apparent Altitudes above the visible Horizon of the Sea.

The Moon's Parallax is the Difference between her Place in the Heavens seen from the Surface of the Earth, and that in which she would be seen from the Centre, which last is called her true Place, and is that which is given directly by Astronomical Tables. On this Account the Moon, except when in the Zenith, always appears lower than her true Altitude; the Quantity of this Depression, called the Moon's Parallax in Altitude, is contained in Table p. 3, 4, 5. and is to be added to all observed Altitudes of the Moon. It is useful in finding the Latitude from the Moon's Meridian Altitude, the apparent Time from the observed Altitude of the Moon at a Distance from the Meridian, and in computing her apparent Altitude from her right Ascension and Declination, the Hour being given; but in this last Case the true Altitude being first found, the Parallax must be subtracted from it to obtain the apparent Altitude.

It

It is likewife ufeful in computing the fecond Correction of Parallax delivered in the Preface to the Britifh Mariner's Guide, which may alfo be confulted for the Application of the abovementioned Ufes of the Moon's Parallax. Mr. Lyon's Tab. IV. of Parallax, defigned for facilitating the Computation of the fecond Correction of Parallax, requires alfo the Table of the Ufe of the Moon's Parallaxes.

The Table, p. 6, 7, and 8, ferves to turn Degrees and Minutes of the Equator into Time, and the contrary; it is of frequent Ufe, as has been fhewn already in the Explication of the Ephemeris. It is alfo ufeful to find the true Difference of Longitude between Greenwich and any Place, from the Difference of Meridians found in Time by the Obfervation of the Moon's Diftance from the Sun or a Star, as will be explained hereafter.

Page 9th contains the Longitudes and Latitudes of 19 of the brighteft Stars and neareft the Ecliptic, being fuch as are moft proper to take the Moon's Diftance from for finding the Longitude at Sea; and therefore it would be better in general not to ufe any others. The 10 marked with Afterifks are the only ones made ufe of in the Diftances of the Ephemeris. This Table is derived from a larger Table of 40 Stars communicated to me by Mr. Gael Morris, deduced from Dr. Bradley's Obfervations, and adapted to the Year 1760. The Longitude of the Stars in this Table being adapted to the Beginning of the Year 1767, muft be increafed by the proportional Part of 50″¼, the annual Variation, for any Day of the Year, according to the Number of Days from the Beginning of the Year, which may be found in the laft Column of the mean Motion of the Sun for the Days of the Month, p. 13—18 in Mayer's Tables: They muft alfo be increafed at the Rate of 50″¼ for every Year after 1767; and muft further be corrected by the Number of Seconds taken out of the following Table, intitled, Table to find the Aberration of a Zodiacal Star in Longitude, communicated alfo by Mr. Gael Morris. To find the Argument or Number for entering this Table with, fubftract the Longitude of the Star from the Longitude of the Sun, borrowing 12 Signs if neceffary;

fary;

fary; where note, that the Character + affixed to the Sign of the Argument, shews that the Number of Seconds is to be added; and — shews that it is to be subtracted; and when the Number of Signs is found at the Bottom of the Table, the Degrees are to be looked for to the right Hand of the Table. The Aberration of Light is an apparent Motion to which all the fixed Stars are subject, the Period of which is completed in a Year. It was first discovered by Dr. Bradley in the Year 1727, and shewn by him to arise from the successive Propagation of Light, and the Motion of the Earth in its Orbit, compounded together. Lastly, the Longitudes of the Stars must be corrected by the Equation of the Equinoctial Points, which is set down for every three Months at the Beginning of the Ephemeris, whence it may be taken at Sight, and applied according to its Sign. This Equation arises from the Nutation of the Earth's Axis, which is owing to the Action of the Moon upon the protuberant Parts of the Earth about the Equator, combined with the Inclination of the Moon's Orbit to the Ecliptic, and the entire Revolution of its Nodes in 18½ Years. This was also discovered by Dr. Bradley, by the like Observations by which he found the Aberration, continued for a Series of twenty Years.

The Moon's Velocity of Access or Recess being greatest with Respect to a Star posited near the same Parallel of Latitude, it is proper to chuse one out of the 24 Stars contained in Table Page 9th, as near this Situation as possible, from which to observe the Moon's Distance for finding the Longitude at Sea: For, if a Star be taken from which the Moon varies her Distance too slowly, the unavoidable Errors of Observation will produce a proportionably greater Error in the Result. The two following Tables, intitled, A Table for chusing proper Stars for observing the Moons Distance from, and a particular Table of Limits for, α Aquilæ, are designed for this Purpose, and were accordingly used for chusing the proper Stars for the Moon's Distances in the Ephemeris. The Use of

the

the Tables is this, the Difference of the Latitudes of the
Moon and Star, if of the same Denomination; viz. both
North or both South, or their Sum, if of contrary Denominations, or one North and the other South, being
found in the first Column, the Difference of the Longitudes of the Moon and Star should not be less than is
shewn against it in the second Column: Or, the Difference of Longitude being found in the second Column, the
Difference or Sum of the Latitudes should not be greater
than is shewn in the first Column: I have adapted the
first Table, so that the Velocity of the Moon's Accefs to or
Recefs from a Star may be never less than Seven Eighths
of her proper Motion; but, in order to take in fo fine
a Star as a Aquilæ, in fome Cafes where there may be a
Defect of other bright Stars proper for the Purpofe, I
have extended the Limits a little further in the fecond
Table, yet fo that the Velocity of the Moon's Accefs to, or
Recefs from, a Aquilæ, may never be lefs than $\frac{3}{4}$ of her
proper Motion.

The Ufe of the following Table of Corrections of the
Moon's Longitude and Latitude found by even Proportion
from the Ephemeris, on account of the fecond Differences
of the Motion in twelve Hours, has been fhewn in the Explanation of the Ephemeris, under the Articles of the
Moon's Longitude and Latitude.

The next Table of the right Afcenfions and Declinations of the principal fixed Stars, is ufeful for finding the
Time and the Latitude by Altitudes taken in the Night;
alfo for computing the Altitude of a Star from which the
Moon's Diftance was obferved, in cafe it was not obferved.
The Method of finding the Time from the obferved Altitude of a Star will be fhewn in the Precepts for finding the
Longitude at Sea by the Help of the Ephemeris. It is alfo
fhewn in the Britifh Mariner's Guide, Chap. iii. Page 19,
which alfo confult at Page 57 and 92, for the two other
Ufes of the Catalogue of right Afcenfions and Declinations,
mentioned above.

If the right Afcenfions of the Stars are required for any Year after 1767, the right Afcenfions in the Table muft be increafed In proportion to the Number of Years after 1767, according to the Increafe of right Afcenfion in Ten Years fet down in Column the fourth: In like manner the Declinations muft be corrected according to the Variation of Declination into Years, fet. down in the laft Column, the Sign + denoting when the Correction is to be added, and the Sign — when it is to be fubftracted.

If the right Afcenfion and Declination are required for any Year before 1767, they are found by diminifhing the right Afcenfion contained in the Table, according to the Number of Years. which precede 1767, and by applying the Correction of Declination with a contrary Sign to that fhewn by the Table.

This Table, as well as the Table of Multipliers, p. 14, is taken from the Britifh Mariner's Guide, which confult at Page 49 for the Ufe of the latter Table.

The following Table of the Depreffion or Dip of the Horizon of the Sea is more correct than the common Tables, the Numbers in it being One Tenth Part lefs than in them. This Correction is owing to the Refraction of the Rays of Light in paffing from the Horizon through the Air to the Eye; and I find it confirmed by Experiment, as well as by Theory. All Altitudes taken from the apparent Horizon of the Sea are to be leffened by the Numbers taken out of this Table, according to the Height of the Eye above the Sea.

The Tables of the right Afcenfions, Declinations, Longitudes, and Latitudes of 21 principal fixed Stars, deduced from Dr. Bradley's Obfervations, were communicated by the Reverend Mr. Hornfby, Savilian Profeffor of Aftronomy at Oxford. They may be prefumed to be very exact, being fettled from Ten Years Obfervations, made between the Years 1750 and 1760; and are fit for the nicer Inquiries of Aftronomy. The Longitudes and Latitudes of 24 Stars, contained p. 9, are moftly the fame with thefe p. 16. to the neareft Second, being both carried on

from

from the fame Settlement of the Stars made to the Beginning of the Year 1760.

Next follow Mr. Lyon's Tables and Rules, and Mr.
Dunthorne's Tables, for correcting the apparent Diftance
of the Moon from the Sun or a fixed Star, on account of
Refraction and Parallax, the Explanation and Ufe of which,
with Examples, is immediately fubjoined to the Tables
themfelves.

EXPLICATION

EXPLICATION and USE

OF THE

T A B L E

OF

PROPORTIONAL LOGARITHMS.

THE apparent Diſtance of the Moon from the Sun, or a fixed Star, being carefully obſerved and reduced to the true Diſtance by the preceding Tables and Rules, it is manifeſt that the Problem of finding the Longitude from Greenwich is reducible to this, to find the apparent Time of the Obſervation by the Meridian of Greenwich; for, the Time at the Ship being given, the Difference of theſe Two is the Difference of Longitude in Time. The apparent Time at Greenwich is to be found by comparing the obſerved Diſtance reduced with the Diſtances of the Moon from the Sun or the ſame Star ſet down in the Ephemeris, for every Three Hours by the Meridian of Greenwich, and making the following Proportion; as the Variation of the Moon's Diſtance in three Hours by the Ephemeris : is to the Difference of the reduced Diſtance, and the next preceding Diſtance in the Ephemeris : : ſo is three Hours or 180' : to the Number of Hours, Minutes, &c. which added to the Hour ſtanding at the Top of the Column above the next preceding Diſtance, gives the apparent Time of the Obſervation by the Meridian of Greenwich.

I have

I have calculated the Table of proportional Logarithms, as I term them, principally in order to render the Operation of this Proportion more eafy. It is analogous to a Table of Logiftical Logarithms, but three times as large, being continued up to 3° or Three Hours, on account of the Diftances being fet down in the Ephemeris at fuch Intervals. Nothing more is to be done than to enter the Table with the Variation of Diftance in Three Hours, looking for the Degree and Minute at Top, and Seconds on the Side, and take out the proportional Logarithm; in like manner to take out the proportional Logarithm of the Difference of the reduced Diftance and next preceding Diftance in the Ephemeris; the Difference of thefe Two Logarithms will be the proportional Logarithm of the Hours, Minutes, and Seconds, which being found in the Table, and added to the Hour of the next preceding Diftance, gives the apparent Time by the Meridian at Greenwich. This will be further explained and illuftrated by an Example among the Precepts for finding the Longitude at Sea by the Help of the Ephemeris.

The Table of proportional Logarithms is alfo very ufeful in facilitating the Computation of the Effeft of Parallax upon the Moon's Diftance from a Star by Mr. Lyon's Rules, as has been fhewn in their Explication at the End of his Tables.

Whenever any Propórtion is to be worked, and all the Three given Terms are Sexagefimals, that is to fay, Degrees, Minutes, and Seconds, Hours, Minutes, and Seconds, &c. the Anfwer is readily found by adding the proportional Logarithms of the fecond and third Terms, together with the arithmetical Complement of the proportional Logarithm of the firft Term; the Sum will be the proportional Logarithm of the fourth Term required. If one of the given Terms is Three Degrees, or Three Hours, whofe proportional Logarithm is o, the Refult is had by only adding Two Numbers together, as in the Cafe firft mentioned of Finding the Time at Greenwich from the reduced Diftance by the Help of the Ephemeris.

T

If

If the Two firſt Terms in the Proportion are common Numbers, and the Third a Sexageſimal, add the proportional Logarithm of the Third Term to the common Logarithm of the Firſt Term, and the arithmetical Complement of the common Logarithm of the Second Term, the Sum will be the proportional Logarithm of the Fourth Term required.

Or, if the Two Firſt Terms are Sexageſimals, and the Third a common Number to the common Logarithm of the Third Term, add the proportional Logarithm of the Firſt Term, and the arithmetical Complement of the proportional Logarithm of the Second Term, the Sum will be the common Logarithm of the Number required.

To multiply a Sexageſimal by any common Number, or by a Sine, Tangent, &c. to the proportional Logarithm of the Sexageſimal, add the arithmetical Complement of the Logarithm of the common Number, or of the logarithmic Sine, Tangent, &c. the Sum will be the proportional Logarithm of the Product required.

The Diviſion of a Sexageſimal will be performed by adding together the proportional Logarithm of the Sexageſimal and the common Logarithm of the Diviſor, the Sum will be the proportional Logarithm of the Quotient required.

The proportional Logarithms were found by ſubſtracting the Logarithm of any Number of Seconds from 4.03342, the Logarithm of 10800, the Number of Seconds contained in 3° or Three Hours.

I ſhall now ſhew, and illuſtrate by an Example, the Manner in which the Moon's Longitude or Latitude may be readily found from the Ephemeris by the Help of this Table. Take half the Increaſe of the Moon's Longitude in Twelve Hours, or the Motion in Six Hours, and again take its Half or the Motion in Three Hours: To the proportional Logarithm of the Moon's Motion in Three Hours, add the proportional Logarithm of the Exceſs of the Time, reckoned from Noon or Midnight, above Three, Six, or Nine Hours, which ever is the next below it, the Sum will

will be the proportional Logarithm of the Degree, Minutes, and Seconds, which added to the Moon's Longitude at the preceding Noon or Midnight, together with the Motion in Three Hours, Six Hours, or the Motions in Three Hours and Six Hours, taken together, gives the Moon's Longitude at the given Time by even Proportion: This must be corrected on account of the Second Differences In the Manner shewn in treating of the Article of the Moon's Longitude.

EXAMPLE.

Let it be required to find the Moon's Longitude and Latitude July 16th 1767, at 16 H. 22 M. 16 S. by the Help of the Ephemeris. July 16th at 12 H. the Moon's Longitude is 0. S. 6°. 40'. 25'. and July 17th at Noon, 0. S. 13°. 47'. 48''. the Difference 7°. 7'. 23''. is the Moon's Motion in 12 Hours; its Half, or 3°. 33'. 41''½. is the Motion In 6 Hours; and its Half again, or 1°. 46'. 51''. is the Motion in 3 Hours. The Time reckoned from Midnight, is, 4 H. 22'. 16'. from which subtracting 3 Hours, there remains 1 H. 22 M. 16 S. Now to 0.2265 the proportional Logarithm of 1°. 46'. 51''. adding 0.3400, the proportional Logarithm of 1 H. 22 M. 16 S. the Sum 0.5665 is the proportional Logarithm of 0°. 48'. 50''. which, together with 1°. 46'. 51''. being added to 0 S. 6°. 40'. 25''. gives 0 S. 9°. 16'. 6''. the Moon's Longitude found by even Proportion : To which add 25''. on account of the Second Differences, and the true Longitude of the Moon will be 0 S. 9°. 16'. 31''. In like manner, to find the Moon's Latitude at the same Time, July 16th at Midnight, by the Ephemeris, it is 4°. 49'. 36''. N. and July 17th at Noon 5°. 3'. 26''. N. Therefore the Motion in 12 Hours is 13'. 50''. and in 3 Hours is 3'. 27''. whose proportional Logarithm is 1.7175, which added to 0.3400, the proportional Logarithm of 1 H. 22'. 16''. gives 2.0575, the proportional Logarithm of 1'. 34''. which added to 3'. 27''. gives 5'. 1''. but this must be corrected by adding

33″. the Correction of Second Differences, whence the proportional Part corrected, is 5′. 1″. + 33″. = 5′. 34″. which added to 4°. 49′. 36″. gives 4°. 55′. 10″. N. the Moon's Latitude correct. See the Example, p. 33—37. under the Explanation of the Article of the Moon's Longitude and Latitude.

INSTRUC-

INSTRUCTIONS

FOR

Finding the Longitude at SEA,

By the Help of the Ephemeris,

ARTICLE I.

Concerning the Instruments and Observations,

THE Observer must be furnished with a good Hadley's Quadrant, and a Watch that can be depended upon for keeping Time within a Minute for six Hours. But it will be more convenient if the Instrument be made a Sextant, in which Case it will measure 120°, for the Sake of observing the Moon's Distance from the Sun, for Two or Three Days after the first and before the last Quarter. The Instrument will be still more fit for the Purpose, if it be furnished with a Screw to move the Index gradually in measuring the Moon's Distance from the Sun or Star; an additional dark Glass, lighter than the common ones, to take off the Glare of the Moon's Light in observing her Distance from a fixed Star, and a small Telescope magnifying Three or Four Times to render the Contact of the Star with the Moon's Limb more discernible. A magnifying Glass of 1½ or Two Inches Focus will assist the Observer to read off his Observation with greater Ease and Certainty.

The greatest Care must be taken in having the Quadrant carefully adjusted before the Observation, or, which I
should

should rather advise, in examining the Error of the Adjustment, for it is liable to alter, and allowing for it. The Method of doing it is this; turn the Index of the Quadrant till the Horizon of the Sea, or the Moon, or any other proper Object appears as One, by the Union of the reflected Image with the Object seen directly; then the Number of Minutes by which o on the Index differs from o on the Arch is the Error of Adjustment. If o on the Index stands advanced upon the Quadrant before, or to the left Hand of o on the Arch, that Number of Minutes is to be subtracted from all Observations; but if it stands off the Arch behind, or to the right Hand of o on the Arch, it must be added to the Observations. But the Sun himself is incomparably the best Object for this Purpose: Either the Two Suns may be brought into One, or, which is a still better Method, the Sun's Diameter may be measured twice, with the Index placed alternately before and behind the Beginning of the Divisions: Half the Difference of these Two Measures will be the Correction of the Adjustment, which must be added or subtracted from all Observations, as the Diameter measured with the Index upon the Arch, that is to say, before or to the left Hand of the Beginning of the Divisions is less or greater than the Diameter measured with the Index off the Arch, behind, or to the right Hand of the Beginning of the Divisions. Thus, suppose I had measured the Sun's Diameter with the Index upon the Arch, or to the left Hand of the Beginning of the Divisions, to be 30′, and the contrary Way to be 33′; I should conclude that the Correction of Adjustment is 1½, or Half the Difference 3′, additive to the Observations. In the Practice of this Method the Telescope must be used, and a dark Glass must be applied at the Eye, or at least on the hither Side of the little speculum, to darken both Suns at once. It will also be convenient to provide an Umbrella of Pasteboard, about Six Inches square, with a Hole in the Middle to receive the Telescope, in order to defend the Eyes from the direct Light of the Sun, as well as from the ambient

ambient Brightness of the Sky, which would otherwise render this Practice in many Cases too painful and difficult.

It will conduce to greater Exactness to take Two or
Three Measures of the Sun's Diameter each Way, Half
the Difference of the Means each Way will be the Correction of the Adiustment, to be applied as before. Thus I
have often assured myself of the exact Quantity of Correction of my Quadrant within a Quarter of a Minute.

There is another Adjustment of the Quadrant, which
is not commonly regarded so much as it ought to be, that
of setting the little Speculum parallel to the great one by
the Screws on the Fore-part of the Instrument. The Manner of doing it is this; hold the Plane of the Quadrant
parallel to the Horizon, and the Index being brought near
to o, if the Horizon of the Sea seen by Reflection in the
little Speculum is higher than the direct Horizon seen by
the Side of it, unscrew one the nearest Screw a little, and
screw up the opposite one till the direct and reflected Ho-
rizons agree. On the contrary, if the reflected Horizon is
lower than the true one, unscrew the Screw farthest from
you, and screw up the nearest one; and take care to leave
the Screws both tight, by screwing them up equally if
they are slack. If this Adjustment is not above 4'. or 5'.
erroneous, it will not be necessary to correct it; and it will
probably never err more, unless the Instrument meets with
some Accident. But for the Sake of Caution it will be
proper to examine it from time to time.

The Observer being now assured of the Adjustment of
his Quadrant, or the exact Correction of it, may proceed
safely to the necessary Observations for ascertaining the
Longitude. The first Observation to be made, is that of
the Altitude of the Sun or some Bright Star, if the Horizon
be fair enough; for computing the apparent Time at the
Ship; and correcting the Watch by which the other Observations are to be made. These Altitudes must not be
taken nearer to the Meridian than Three or Four Points;
but the nearer they are taken due East or West the better,
provided the Objects be not less than 5°. high. The next
Observation to be made is that of the Distance of a Star

from

from the Moon's enlightened Limb, or the Diftance of the neareft Limbs of the Sun and Moon. The Two other requifite Obfervations are the Altitudes of the Moon and Star, or of the Moon and Sun, to be taken by Two Affiftants at the very Inftant, or at the utmoft within a Minute of the Time at which the principal Obferver gives Notice of the completing his Obfervation of the Diftance of the Moon from the Sun or Star. At the fame Inftant, or at the utmoft within a Quarter of a Minute, and before the Obfervers attempt to read off the Degrees and Minutes from their Quadrants; fomebody muft note the Hour, Minute, and Quarter Part of a Minute (if there is no fecond Hand) of the Watch ufed in taking the Sun or Star's Altitude for computing the Time; and the Obfervations requifite for afcertaining the Longitude are completed.

If the Moon's Diftance be taken from the Sun, and the Sun be not nearer to the Meridian than Three Points, and his Altitude be well taken within 15". or 20". of the Obfervation of the Diftance, this Altitude will ferve to compute the apparent Time, without requiring the Ufe of the feparate Obfervation firft mentioned, except it be ufed by way of Confirmation and Check both upon Obfervation and Calculation.

In taking the Moon's Diftance from the Sun, the Obferver muft look at the Moon directly through the unfoiled Part of the little Speculum, and obferve the Sun by Reflection, letting down One of the dark Glaffes ufed in taking his Meridian Altitude. In taking the Moon's Diftance from a Star, he muft look at the Star directly, and fee the Moon by Reflection, ufing the dark Glafs that is lighter than the Reft; and defigned for this particular Purpofe. The Plane of the Quadrant muft be always made to pafs through the Two Objects whofe Diftance is to be obferved, and muft be put into various Pofitions according to the Situations of the Objects, which will be rendered familiar by a little Experience.

In order to attain the greater Degree of Exactnefs, it will be better to repeat the Obfervations till at leaft Three Diftances and their correfponding Altitudes are obtained ;

but the more that are taken the better. The Sum of the Diſtances divided by the Number of them is the mean Diſtance: in like manner the mean Altitudes, and the mean Time by the Watch are obtained; which then are to be uſed as a ſingle Obſervation would be, only they may be relied upon with greater Aſſurance. But theſe Obſervations muſt be all included within the Space of Half an Hour.

The Manner of finding the Star, whoſe Diſtance from the Moon is ſet down in the Ephemeris, has been mentioned among the Uſes of the Diſtances contained in the Ephemeris.

Whoever would ſee more concerning the neceſſary Inſtruments and Obſervations, may conſult the Two firſt Chapters of the Britiſh Mariner's Guide, from which moſt of the foregoing Inſtructions are borrowed.

ARTICLE. II.

To compute the apparent Time from the obſerved Altitude of the Sun or a known Star, and thence to find the apparent Time of the Obſervation of the Diſtance of the Moon from the Sun or a Star.

From the obſerved Altitude of the Sun's lower Limb, ſubſtract the Sum of the Dip and Refraction, taken out of Page 2d and 10th of requiſite Tables, and to the Remainder add 16'. for the Sun's Semidiameter (or if you have a mind to be more exact, make uſe of the Sun's Semidiameter, ſhewn Page 3d of the Month in the Ephemeris) and you have the true Altitude of the Sun's Centre. If the Sun's upper Limb was obſerved, his Semidiameter muſt be ſubſtracted inſtead of being added. If the Altitude of his Centre was taken, it is only neceſſary to ſubſtract the Sum of the Dip and Refraction. Subſtract the true. Altitude of the Sun thus found from 90º. and you have his true Zenith Diſtance.

U

The

The Sun's Declination is to be found from the Ephemeris, Page 2d of the Month; but being there set down for apparent Noon at Greenwich, Proportion must be made to find what It should be at the given Time reduced to the Meridian of Greenwich. Turn your Longitude by Account from London or Greenwich Into Time, by Table Page 6, 7, and 8, which add to, or subtract from, the Time at the Ship, estimated nearly according as you are to the West or to the East of Greenwich: This gives the Time at Greenwich. Then say, as 24 H. is to this Time, so is the daily Variation of the Sun's Declination in the Ephemeris to a Number of Minutes, &c. which added to, or subtracted from, the Sun's Declination, at the preceding Noon in the Ephemeris, according as his Declination is increasing or decreasing, gives his true Declination required. Note, that the Sun's Declination may be found in the same Manner for computing his Azimuth, to compare with his observed Azimuth in order to find the Variation of the Compass. The Sun's Declination, if of the same Denomination with the Latitude of the Place (viz. both North or both South) must be subtracted from 90°. but if of a contrary Denomination to the Latitude of the Place (viz. one North and the other South) must be added to 90°. the Sum or Difference is the Distance of the Sun from the Pole of the World which is above the Horizon. Find also the Latitude of the Ship, at the Time of taking the Altitude of the Sun, by allowing for the Ship's Run from the Latitude determined at the nearest Meridian Observation before or after: The Complement to 90°. is the Co-Latitude.

Now add together the Zenith Distance, Polar Distance, and Co-Latitude, and take Half the Sum, and the Difference between the Half Sum and the Zenith Distance: Then add the Sines of the Half Sum and the said Difference, together with the arithmetical Complements of the Sines of the Polar Distance and Co-Latitude, Half the Sum of these Four Logarithms is the Cosine of Half the horary Angle; which therefore doubled gives the horary Angle or true Distance of the Sun from the Meridian. This being turned into

Time

Time by Table Page 6, 7, and 8, gives the apparent Time
if it be Afternoon; but, if it be Forenoon, the Complement to Twenty-four Hours is the apparent Time reckoned
from the preceding Noon. Five Places of Logarithms, be-
fides the Index, will be fufficient for this Computation.

The Difference between the apparent Time thus found,
and the Time fhewn by the Watch at the Inftant of taking
the Altitude, fhews how much the Watch is too faft or too
flow; which Difference being applied as a Correction to
the Time fhewn by the Watch when the Diftance of the
Moon from the Sun or Star was taken, being added there-
to, if the Watch is too flow, or fubftracted therefrom, if
the Watch is too faft, gives the apparent Time of the Obfervation of the Diftance.

EXAMPLE.

Suppofe the apparent Altitude of the Sun's lower Limb
above the Horizon of the Sea fhould be obferved April 4th
1767 to be 47°. 13'. the Height of the Eye above the Sea
being 18 Feet; the Latitude of the Ship at the fame time
corrected for the Run from the preceding Noon, being 16°.
24'. North, the Longitude to the fame Time by the Ship's
Reckoning 43°. 37'. Weft of Greenwich, and the Time
at the Ship eftimated nearly 2 H. 41 M. It is required
to find the apparent Time?

Obferved Altitude of the Sun's lower Limb	47°.	13'.
Sum of Dip 4'. and Refract. 1'. fubftract		5
	47.	8
Sun's Semidiameter, add — —		16
True Atitude of Sun's Centre —	47.	24
Subftract from — — —	90	
True Zenith Diftance of the Sun —	42°.	36'.

Time

		H.	M.
Time from Noon eftimated nearly ——		2	41
Long. W. of Greenwich *per* Acc. 43° 37′. =	2	54	28

App. Time at Greenwich nearly —·   5   35

The Sun's Declination April 4th at Noon by the Ephemeris is 5°. 42′. 51″. N. and April 5th is 6°. 5′. 39″. N. The Difference or daily Increafe is 22′. 48″. Say then, as 24 H. is to 5 H. 35′. fo is 22′. 48″. to 5′. 18″. which added to 5°. 42′51. ″. the Sun's Declination in the Ephemeris for the preceding Noon, gives 5°. 48′. 9″. N. the Sun's true Declination at the Time required, or rejecting the Seconds, 5°. 48′. N. the Complement to 90°. (becaufe the Latitude and Declination are of the fame Denomination) gives 84°. 12′. for the Sun's true Diftance from the North or elevated Pole.

The Latitude of the Ship carried on by Account from the preceding Noon is 16°. 24′. N. the Complement of which to 90°. is 73°. 36′. the Co-Latitude.

Zenith Dift. of the Sun	42. 36		
Polar Dift. of the Sun	84. 12	Ar. Com. Sine	0.00223
Co-Latitude — —	73. 36	Ar. Com. Sine	0.01804

Sum — — —	200. 24		
Half Sum — —	100. 12	Sine	9.99308
Half Sum — Zenith Dift.	57. 36	Sine	9.92651

Sum of 4 Logarithms	—	—	19.93986
Half Sum Cofine of — 21°. 4′ 1,			9.96993
2			

Sun's horary Angle — 42. 9
                  H. M. S.
Therefore app. Time   2. 48. 36.

Suppofe the Watch at the Sun's Altitude fhewed 2 H. 56′. 48″. the Difference is 8′. 12″. by which the Watch is too faft for apparent Time. Now if the Moon's Diftance from

from a Star was obferved afterwards at 8 H. 27'. 18". by the Watch, fubftraft 8'. 12". and the apparent Time of Obfervation of the Diftance is 8 H. 19'. 6".

To find the apparent Time from the obferved Altitude of a known fixed Star.

The obferved Altitude of the Star above the Horizon of the Sea muft be leffened by the Sum of the Dip and Refraction, the Remainder is the true Altitude of the Star, and the Complement to 90°. is the Zenith Diftance. The Declinations of the principal fixed Stars are contained in Table p. 12 and 13, which muft be corrected for the Increafe or Decreafe for any Year after 1767, according to the Variation in Ten Years fet down in the laft Column. The Declination fubftracted from or added to 90°. as the Declination of the Star and the Latitude of the Ship are of the fame or contrary Denominations, gives the polar Diftance of the Star.

The Zenith Diftance and Polar Diftance of the Star and Co-Latitude being found, the Diftance of the Star from the Meridian is found by the very fame Method or Procefs of Logarithms as was before fhewn for finding the Time by the Sun. Then fubftract the Diftance of the Star Eaft of the Meridian from its Right Afcenfion (found by Table p. 12 and 13. corrected for any Number of Years after 1767, according to the Increafe in Ten Years, fet down in 5th Column) or add the Diftance of the Star Weft of the Meridian to the right Afcenfion of the Star, the Difference or Sum is the right Afcenfion of the Midheaven : which turn into Time by Page 6, 7, and 8. From this (borrowing 24 Hours if neceffary) fubftract the Sun's right Afcenfion in Time at the preceding Noon at Greenwich ftanding in the Ephemeris, the Remainder is the apparent Time nearly. To which adding or fubftracting the Longitude of the Ship from Greenwich, turned into Time, according as it is to the Weft or to the Eaft of Greenwich, you will have the apparent Time nearly by the Meridian of Greenwich. Then fay, as 24 H. is to this Time ; fo is the daily Variation of the Sun's right Afcenfion in Time

by

by the Ephemeris, to a Number of Minutes and Seconds; which fubftracted from the apparent Time at the Ship, found nearly above, leaves the apparent Time correct.

EXAMPLE.

Suppofe the Altitude of the Star Procyon above the Horizon of the Sea, fhould be obferved Sept. 7th 1767 in Latitude 7°. 45'. South, Longitude 30°. 10'. Eaft of Greenwich per Account, to be 28°. 16'. the Height of the Eye above the Sea being Eighteen Feet. Required the apparent Time?

Obferved Alt. of Procyon   ——   28°. 16'.
Sum of Dip 4', and Refraction 2'. fubft,   6

True Alt. of Procyon —— — 28. 10
Complement to 90°. or Zenith Diftance   61. 50
Declination of Procyon by Page 12th.   5. 49 N.
Increafed by 90°. is — — — — 95. 49
the Diftance of Procyon from the South or elevated Pole.
The Latitude is — — — 7. 45 S.
Therefore Co-Latitude — — — 82. 15
Zenith Diftance   61°. 50'.
Polar Diftance —   95. 49 Ar. Comp. Sine 0.00224
Co-Latitude —   82. 15 Ar. Comp. Sine 0.00398

Sum — —   239. 54
½ Sum —— 119. 57 Sine ——   9.93775
½ Sum — Zen. Dift. 58. 7 Sine ———   9.92897

Sum of Four Logarithms — —   19.87294
Half Sum, is Cofine of ——   30. 14½   9.93647
     2

Doubled is horary Angle or ⎫
   Diftance of Procyon from the ⎬ 60. 29
   Meridian to the Weft ⎭
Therefore fubftr. from right ⎱ 111. 47
   Afc. of Star Page 12th ⎰

Right Afc. of Mid-heaven   51. 18

Or

	H.	M.	S.
Or, in Time by Page 6th —	3.	25.	12
Subſtract Sun's right Aſc. in Time Sept. 7th at Noon by the Ephemeris — —	11.	3.	7
Apparent Time nearly	16.	22.	5
Long. 30°. 18'. Eaſt of Greenwich in Time, ſubſtract	2.	1.	12
Leaves apparent Time at Greenwich nearly —	14.	20.	53

The Sun's right Aſcenſion in Time, Sept. 7th at Noon
being 11 H. 3'. 7". and Sept. 8th 11 H. 6'. 42". the daily
Variation is 3'. 35". Then ſay as 24 H. is to 14 H. 21
M. ſo is 3'. 35". to 2'. 8". which ſubſtracted from 16 H.
22'. 5". the apparent Time at the Ship found nearly above,
leaves 16 H. 19'. 57". the apparent Time correct.

Note, If the Longitude of the Ship Eaſt of Greenwich
in Time is greater than the apparent Time at the Ship, it
will be neceſſary to borrow 24 Hours, in order to find
the apparent Time at Greenwich ; but if the Longitude
Weſt of Greenwich in Time, added to the apparent Time
at the Ship, makes more than 24 Hours, 24 H. muſt be
ſubſtracted from the Sum to have the apparent Time at
Greenwich : And in the firſt Caſe the Sun's right Aſcenſion in Time muſt be taken out of the Ephemeris for one
Day of the Month leſs than that reckoned at the Ship ;
and in the other Caſe it muſt be taken out for one Day
more. The Sun's right Aſcenſion thus found in either
of theſe Caſes, is to be ſubſtracted from the right Aſcenſion of the Mid-heaven, to find the apparent Time nearly,
which muſt be corrected by the proportional Part of the
Sun's daily Variation of right Aſcenſion, in like manner as
has been ſhewn before.

ARTICLE

ARTICLE III.

To reduce the obferved Diftance of the Moon's
Limb from a Star, or from the Sun's Limb to
the true Diftance of the Centres.

The apparent Time of the Obfervation of the Moon's
Diftance from the Sun or a Star being found by the preceding Article, add to it or fubftract from it the Longitude
from Greenwich by Account turned into Time, according as the Ship is to the Weft or to the Eaft of Greenwich,
and you will have the apparent Time at Greenwich nearly; with which take out of the Ephemeris, from Page
8th of the Month, the Moon's Semidiameter, horizontal
Parallax, and its logiftic Logarithm for 1767, or proportional Logarithm for fubfequent Years; alfo the Sun's Semidiameter, from Page 3d of the Month, if the obferved
Diftance was that of the Moon from the Sun. But the Articles, contained Page 8th of the Month, being fet down
in the Ephemeris only for Noon and Midnight, it will be
neceffary to make Proportion to find them for any inter-
mediate Time.

Take the Difference of the Two Semidiameters of the
Moon, &c. ftanding in the Ephemeris againft the Noon
and Midnight, which immediately precede and follow the
given Time reduced to the Meridian of Greenwich, and
you have the Variation of the Semidiameter, &c. in 12
Hours: Then fay, as 12 H. is to the apparent Time reduced to Greenwich, reckoned from the preceding Noon
or Midnight, fo is the Variation of the Semidiameter, &c.
in 12 Hours, to the proportional Part required; which
added to the Moon's Semidiameter, &c. at the preceding
Noon or Midnight, if it is increafing. or fubftracted from
it, if it is decreafing, gives the Moon's Semidiameter, &c.
at the given Time. The Moon's Semidiameter thus found
is to be augmented according to her Altitude, as follows,
to obtain the apparent Semidiameter.

$\mathcal{D}$'s Alt.

☽ Alt. 5°.10°.15.20°.25°.30°.35°.40°.45°.50°.55°.60°.65°.70°.75 & above
Inr. ☽ 1″. 3″. 4″. 6″. 7″. 8″. 9″. 10″.11″.12″.13″.14″.15″.15″.16″.
Semid.

to the logiſtic Logarithm of the Moon's horizontal Parallax,
found from the Ephemeris of the Year 1767, add the con-
ſtant Logarithm 0.4771, rejecting 1, when it ariſes in the
Place of the Index, and you will have the proportional Logarithm of the horizontal Parallax, which muſt always have
the Cypher o prefixed in the Place of the Index.

E X A M P L E.

Suppoſe it was required to find the Moon's apparent Semidiameter, horizontal Parallax, and proportional Logarithm of the fame, Nov. 5th 1767, at 10 H. 27 M.
apparent Time in the Longitude 78°. 13′. Weſt of Greenwich. The Longitude turned into Time is 5 H. 12 M.
52 S; which added to 10 H. 27 M. (becauſe the Longitude
is Weſt) gives 15 H. 40 M. (to the neareſt Minute) for
the apparent Time at Greenwich, or 3 H. 40 M. after
Midnight. Now the Moon's Semidiameter Nov. 5th at
Midnight, by the Ephemeris, is 16′. 28″. and Nov. 6th
at Noon, is 16′. 24″. therefore the Decreaſe in 12 Hours
is 4″. Then ſay, as 12 H. is to 3 H. 40 M. ſo is 4″. to
1″. which ſubſtracted from 16′. 28″. becauſe (Moon's Semidiameter is decreaſing) leaves 16′. 27″. the Moon's horizontal Semidiameter at the given Time. In like manner
the Moon's horizontal Parallax will be found 60′. 25″.
— 5″. = 60′. 20″. and the logiſtic Logarithm of the
fame 9970 + 5 = 9975, becauſe it is increaſing. Add
0.4771, and the Sum, rejecting 1 in the Place of the Index,
is 0.4746, the proportional Logarithm of the Moon's horizontal Parallax. Suppoſe the Moon's Altitude to be 52°.
the Increaſe of her Semidiameter, anſwering to this Altitude ſhewn above, is 12″. which added to her horizontal
Semidiameter 16. 27″. found above, gives her apparent
Semidiameter 16 39″.

Now add the Moon's apparent Semidiameter, juſt found,
to the obſerved Diſtance of the Moon's Limb from a Star.

X   If

If it was the Limb neareft the Star; but fubftract the
Moon's apparent Semidiameter from the obferved Diftance,
if the Limb obferved was that furtheft from the Star, and
you will have the apparent Diftance of the Moon's Centre
from the Star. But to the obferved Diftance of the Sun
and Moon's neareft Limbs add the Sum of the apparent Semidiameters of the Moon and Sun, and you will have the
apparent Diftance of their Centres.

Subftract the Quantity of the Dip of the Horizon of the
Sea from the obferved Altitude of the Star; and add 16′.
leffened by the Dip to the obferved Altitude of the Sun or
Moon's lower Limb; but fubftract the Sum of the Dip
and 16′. from the obferved Altitude of the Moon's upper
Limb, and you will have the apparent Altitudes of the
Moon and Star or Sun.

Laftly, with thefe Altitudes, and the apparent Diftance
of the Moon's Centre from a Star, or the Sun's Centre
found before, and the Moon's horizontal Parallax, or its
proportional Logarithm, found in Manner fhewn above,
compute the Corrections neceffary to be made on account of
Refraction and Parallax, either by Mr. Lyons' or Mr. Dunthorne's Tables, being Part of this Work, in the Manner
explained immediately after the Tables themfelves; which
being applied, according to thofe Directions, to the apparent Diftance of the Moon's Centre from a Star, or the
Sun's Centre, will give the true or reduced Diftance of the
Moon from the Star or Sun.

ARTICLE IV. and laft.

To find the Longitude from the obferved Diftance
reduced, by the Help of the Ephemeris.

Take the Difference of the next lefs and next greater
Diftances, ftanding in the Ephemeris, then the reduced
Diftance, gives the Variation of Diftance in Three Hours.

Take the Difference between the reduced Diftance and
the next preceding Diftance in the Ephemeris, namely, the
next

next lefs Diftance when it is increafing, or next greater Diftance if it is decreafing; this call the Difference of Diftance.

Subftract the proportional Logarithm of the Variation of
Diftance in Three Hours from the proportional Logarithm
of the Difference of Diftance, gives the proportional Logarithm of the Hour with Minutes and Seconds; which
added to the Hour at Greenwich of the next preceding
Diftance, gives the true Time of the Obfervation of the
Moon's Diftance from the Sun or Star by the Meridian of
Greenwich; the Difference between this and the Time of
the Obfervation at the Ship is the Longitude of the Ship
from the Meridian of Greenwich in Time; and is Eaft or
Weft, as the Time at the Ship is greater or lefs than that
at Greenwich. This is to be turned into Degrees and Minutes of Longitude, at the Rate of One Hour to 15°, or
more briefly by the Table Page 6, 7, and 8th.

EXAMPLE.

Suppofe the Moon's Diftance from Regulus to the Eaft
of her obferved and reduced Diftance fhould be found Jan.
13th 1767 at 10 H. 27 M. 13 S. apparent Time at Sea to
be 46°. 32'. 24''.. Look in the Ephemeris againft Jan.
13th for the next greater and next lefs Diftances than the reduced Diftance, and you will find 47°. 20'. 18''. at 6 H. and
45°. 48'. 52''. at 9 Hours by the Meridian of Greenwich:
the Difference of thefe is 1°. 31'. 26''. which is the Variation of Diftance in Three Hours, whofe proportional
Logarithm is 2942.

Take alfo the Difference of the reduced Diftance 46°.
32'. 24''. and 47°. 20'. 18''. the next Diftance preceding the
reduced one (namely the next greater, becaufe the Diftance
in this Cafe is decreafing) and you will have 47'. 54''. for
the Difference of Diftance, whofe proportional Logarithm
is 5749; from which fubftract 2942 the proportional Logarithm of the Variation of Diftance in Three Hours found
above, and there will remain 2807, the proportional Logarithm of 1 H. 34'. 19''. This added to Six Hours, the
Hour of the next preceding Diftance in the Ephemeris,
gives

gives 7 H. 34'. 19". apparent Time by the Meridian of Greenwich ; the Difference of which and 10 H. 27'. 13", the apparent Time at the Ship, or 2 H. 52'. 54". is the Longitude of the Ship, reckoned from the Meridian of Greenwich in Time, which by the Table, p. 6, 7, and 8, gives the Longitude 43°. 13'. 30". East, becaufe the Time at the Ship is greater than that at Greenwich.

N. B. The Longitude thus found is that of the Ship, when the Altitude of the Sun or Star was taken for regulating the Watch, and not when the Diftance of the Moon from the Star was obferved, unlefs the Altitude made ufe of for computing the Time was made at or very near the Time of the Obfervation of the Diftance.

The Longitude is to be carried on to the following Noon, and fo on from Day to Day, by the Ship's Reckoning in the ufual Manner, until it is again afcertained by fubfequent Obfervations.

END of the Inftructions for finding the Longitude at Sea by the Help of the EPHEMERIS.

Here follow two Examples of the Calculation of the Longitude by the Help of the Ephemeris, one from the Diftance of the Moon from the Sun, and the other from the Diftance of the Moon from a Star.

EXAMPLE I.

Suppofe the following Obfervations fhould be taken at Sea, April 4, 1767.

April 4, 1767. Time by the Watch.	Obferved Dift. of Sun and Moon's neareft Limbs.	Obferved Alt. of Sun's L. L. from Horizon of Sea.	Obferved Alt. of Moon's L.L. from Horizon of Sea.
h. m. s.	° ′ ″	° ′	° ′
4.47.14	73.41.53	22. 50	80. 17
4.50.11	73.43.55	22. 12	80. 36
4.53.26	73.47.33	21. 6	81. 9
Mean of the Time	Mean Diftance	Mean Alt. Sun's ower Limb.	Mean Alt. Moon's ower Limb.
4.50.57	73.44.27	22. 3	80. 41

Suppofe alfo that at 5ʰ. 4′. 38″. by the Watch, a little after the foregoing Obfervations, the apparent Altitude of the Sun's lower Limb above the Horizon of the Sea was obferved 19°. 13′, in order for computing the Time, the Height of the Eye above the Sea being 18 Feet, the Latitude being 34°. 17′ N. and the Longitude by Account 17°. 46′ Weft of Greenwich.

Computation of the apparent Time by Article II.

The Sum of the Dip 4′, and Refraction 3′ or 7′ being fubftracted from the obferved Altitude of the Sun's lower Limb 19°. 13′ leaves 19°. 6′ to which 16′ being added for the Sun's Semidiameter, the true Altitude of the Sun's Centre is 19°. 22′, and his zenith Diftance, 70°. 38′. The Time of this Altitude by the Watch, is 5ʰ. 4′. 38″. and fuppofe the Watch is eftimated to be 20 Minutes too faft for apparent Time, then 20 Minutes fubftracted leaves 4ʰ. 44′. apparent Time eftimated nearly ; to which add 1ʰ. 11′ for the Longitude by Account Weft of Greenwich in Time, and you have 5ʰ. 55′ for the Time at Greenwich eftimated nearly.

Therefore the Sun's Declination by the Ephemeris is 5°. 48′ N. and the polar Diftance 84°. 12′. The Latitude being 34°. 17′ N. the Colatitude is 55°. 43′.

Zenith Diftance of the Sun 70. 38 Ar. com. Sine 0, 00223
Polar Diftance of the Sun 84. 12 Ar. com. Sine 0, 08268
Colatitude · 55. 43

Sum	210. 33	
½ Sum	105. 16	Sine 9, 98440
½ Sum—Zen. Dift.	34. 38	Sine 9, 75459

Sum—4 Logarithms		19, 82410
½ Sum, Cofine of	35. 15	9, 91205
	2	

Sun's horary Angle 70. 30

There

	h	m	s
Therefore apparent Time	4.	42.	o
But Time by Watch was	5.	4.	38
Therefore Watch is too fast		22.	38

Subſtract from 4ᵇ. 50ᵐ. 57ˢ. the Mean of the Times by Watch at the Obſervations of the Diſtance of the Moon from the Sun, and there remains 4ᵇ. 28ᵐ. 19ˢ. apparent Time.

Reduction of the obſerved Diſtance of the Moon from the Sun according to Article III.

The Mean of the 3 obſerved Altitudes of the Sun's lower Limb above the Horizon of the Sea is 22°. 3′. to which adding 16′—4′ or 12′, for the Sun's Semidiameter leſſened by the Dip, and the apparent Altitude of the Sun's Centre is 22°. 15′.

The Mean of the 3 obſerved Altitudes of the Moon's lower Limb is 80°. 41′, to which 12′ being added, the apparent Altitude of the Moon's Centre is 80°. 53′.

The apparent Time at Greenwich eſtimated nearly being 5ᵇ. 55′, the Moon's Semidiameter by the Ephemeris, is 15′. 19″, the Moon's horizontal Parallax 56′. 12″, and the logiſtic Logarithm of the ſame 0283, to which add 0,4771, and the proportional Logarithm is 0,5054. To the Moon's Semidiameter 15′. 19″, adding 16″ for her Altitude 81°, the apparent Semidiameter of the Moon is 15′ 35″ : The Sun's Semidiameter by Page 3d of the Month is 16′.1″; therefore the Sum of the apparent Semidiameters of the Sun and Moon is 31′ 36″; which added to 73°. 44′. 27″, the mean of the 3 obſerved Diſtances of the neareſt Limbs of the Sun and Moon, gives 74°. 16′. 3″ for the apparent Diſtance of the Centres.

The principal Effect of Refraction will be found from Tab. I. of Mr. Lyons 176″, from which 32″ found by Table II. being ſubſtracted, leaves 144″ = 1′.24″. for

the

the Effect of Refraction, which added to 74°. 16'. 3"
gives 74°. 18'. 27". the Diflance cleared of Refraction.
Arches the firft and fecond for Parallax by Mr. Lyon's
Rules are 22'. 5". and 15'. 36"; the Difference 6'. 29".
fubftracted from 74°. 18'. 27". (becaufe Arch the firft is
greateft) leaves 74°. 11'. 58". the reduced Diflance :
Table IV. for Parallax in this Cafe gives 0.
By Mr. Donthorne's Tables the reduced Diflance will
be found 74°. 11'. 59".

Determination of the Longitude from the reduced
Diflance by Article IV.

The Diflance of the Moon from the Sun, April 4, 1767,
ftanding in the Ephemeris next preceding the reduced
Diftance 74°. 11'. 58" is 73°. 1'. 27", at 3ʰ. and the
Diflance next following it at 6ʰ. is 74°. 28'. 50"; therefore the Variation of Diftance in 3 Hours is 1°. 27'. 23",
whofe proportional Logarithm is 0,3138. The Difference
of the reduced Diflance 74°. 11'. 58". and the next preceding Diflance 73° 1'. 27" is 1°. 10'. 31", whofe proportional Logarithms is 0,4070 : The Differenceof thefe two
proportional Logarithms is 0,0932, the proportional Logarithm of 2ʰ. 25'. 14". Add 3ʰ. ftanding over the next
preceding Diftance, and 5ʰ. 25'. 14" is the apparent Time
at Greenwich; but the apparent Time at the Ship was
found before 4ʰ. 28'. 19": The Difference 0ʰ. 56'. 55"
converted into Degrees, &c. by Table Page 6, 7, and 8,
gives 14°. 13'. 45". the Longitude from Greenwich at
the Time when the Altitude of the Sun was taken from
which the Time was computed, and it is Weft becaufe the
Time at the Ship is kaft.

EXAMPLE

Example II.

Sept. 5. 1767. At 4ʰ. 59ᵐ. 32ˢ. by Watch, let the obferved Altitude of the Sun's lower Limb above the Horizon
of the Sea be 10°. 29′. The Height of the Eye above the Sea
12 Feet, the Latitude at the fame Time 17°. 30′ S. and
the Longitude by Account 64°. 32′. Eaſt of Greenwich.

Time by Watch.	Obſerved Diſtance Moon's remoteſt Limb from Sun's	Apparent Altitude Moon's L. L. above Horizon of Sea.	Apparent Altitude [Star above Hu-rizon of Sea.
h. m. s	° ′ ″	° ′	° ′
14. 50. 30	44. 53. 48	15. 22	35. 22
14. 55. 35	44. 50. 49	14. 11	34. 15
15. 00. 0	44. 48. 14	13. 8	33. 17
15. 5. 50	44. 44. 48	11. 46	32. 1
15. 11. 15	44. 41. 37	10. 30	30. 50
Mean of Times	Mean of Diſtan.	Mean Altitude of Moon's L. L.	Mean Altitude of Star.
15. 0. 38	44. 47. 51	12. 59	33. 9

The Sum of the Dip 3′, and Refraction 5 or 8′ ſubſtracted from 10°. 29′. the obſerved Altitude of the
Sun's lower Limb, leaves 10°. 21′, to which 16′ being
added for the Sun's Semidiameter, the true Altitude of
the Sun's Centre is 10°. 37′, and his zenith Diſtance 79°.
23′. The Time at Greenwich eſtimated nearly being
8ʰ. 41′, the Sun's Declination by the Ephemeris is 6°.49′ N.
and his Diſtance from the South or elevated Pole 96°. 49′.
The Latitude being 17°. 30′. the Colatitude is 72°.
30′. Hence the apparent Time will be found 5ʰ. 6′.
16″. But the Time by the Watch was 4ʰ. 59′. 32″.
Therefore the Watch is 6′. 44″ too ſlow. Add this to
15ʰ. 0′. 38″, the Mean of the Times by the Watch at the
Obſervations of the Diſtances, and the apparent Time of
thoſe Obſervations will be 15ʰ. 7ᵐ. 22ˢ.

To

To 12°. 59' the mean Altitude of the Moon's lower
Limb obferved, add 16'—3' or 13' for the Moon's Semidiameter leffened by the Dip, the apparent Altitude
of the Moon's Centre will be 13°. 12'. From 33°. 9'.
the obferved Altitude of the Star, fubftract 3' for the
Dip, and the apparent Altitude of the Star is 33°. 6'.
The Longitude Eaft of Greenwich In Time by Account
4ʰ. 18ᵐ. 8ˢ. (anfwering to 64°. 32') fubftracted from 15ʰ.
7'. 22'', leaves 10ʰ. 49ᵐ. the apparent Time at Greenwich
eftimated nearly. Hence the Moon's Semidiameter, by
the Ephemeris, is 16'. 29'', the horizontal Parallax 60'.
32'', and its logiſtic Logarithm 9962, and confequently
the proportional Logarithm 0,4733. The Increafe of the
Moon's Semidiameter for the Altitude 13° is 4'', whence
the Moon's apparent Semidiameter is 16'. 33'', which fubftracted from 44°. 47'. 51'', the Mean of the obferved Diftances of the Moon's remote Limb from a Pegaſi, leaves
44°. 31'. 18'', the apparent Diftance of the Moon's Centre
from the Star.

Hence the Effect of Refraction will be found by Mr.
Lyons' Tables+1'.50'', and the apparent Diftance cleared
of Refraction 44°. 33'. 8''. Arches the firft and fecond for
Parallax will be found 47'. 6'' and 13'. 56'', and the principal Effect of Parallax or Parallax in Diftance 33'. 8'' to
be fubftracted. The Number correfponding to this by
Table 4th for Parallax is 9'', and that correfponding to
58'1, the Moon's Parallax in Altitude, is 30'', the Difference 21''is the fecond Correction of Parallax to be added;
and 44°. 33'. 8'' — 33'. 8'' + 21'' = 44°. 0'. 21'' the
Diftance reduced. The fame comes out by Mr. Dunthorne's
Tables 44°. 0'. 23''. The next preceding Diftance in the
Ephemeris is 45°. 6'. 40'' at 9ʰ, and the next following
Diftance at 12ʰ. is 43°. 24'. 24''; whence the Difference or
Variation of Diftance in 3 Hours is 1°.42'. 16''. whofe
proportional Logarithm is 2455 : The Difference of the
reduced Diftance 44°. 0'. 22'', and the preceding Diftance
45°. 6'.40'', is 1°. 6'. 18'', whofe proportional Logarithm is 4338 : The Difference of thefe Logarithms is
1883 the proportional Logarithm of 1ʰ. 56'. 40''.

Y   Add

Add 9^h, and the apparent Time at Greenwich is 10^h. 56^m. 40^s. But the apparent Time at the Ship was 15^h. 7^m. 22^s : The Difference 4^h. $10'$. $42''$ converted into Degrees gives 62°. $40'$. $30''$ for the Longitude from Greenwich at the Time when the Altitude of the Sun was taken for computing the Time, and it is East, because the Time at the Ship is greatest.

N.B. An Altitude of the Sun for computing the Time might have been taken early in the Morning after the Observations of the Distance, which would have had the Advantage of the Altitude taken in the Afternoon by being nearer the Time of the Distance, so that there would be Occasion to depend upon the going of the Watch for a less Interval of Time; besides that, the Longitude would be thereby carried on for a longer Time.

F I N I S,

I N D E X

Of the Contents of the NAUTICAL ALMANAC and
ASTRONOMICAL EPHEMERIS, and of TABLES
requisite to be used with the same.

INDEX.

END of the TABLE OF CONTENTS.

ADDITION

TO

Mr. DUNTHORNE's Solution of the Problem for finding the Effect of Refraction and Parallax.

Communicated by Mr. DUNTHORNE.

IF we have the diftance of the Moon from the Sun, inftead of a ftar (and great exactnefs be required) his parallax in altitude, taken from the following table, muft be fubftracted from his refraction, and only the difference ufed in the room of the refraction of a ftar.

A TABLE of the Sun's Parallax in Altitude.

Altitude of the Sun.	Parallax.	Altitude of the Sun.	Parallax.	Altitude of the Sun.	Parallax.
°	''	°	''	°	''
0	9	30	8	60	4
5	9	35	7	65	4
10	9	40	7	70	3
15	9	45	6	75	2
20	8	50	6	80	2
25	8	55	5	85	1
30	8	60	4	90	0

ERRATA of the TABLES requisite to be used with the ASTRONOMICAL and NAUTICAL EPHEMERIS.

Page 17. Mr. Lyon's first Table for refraction.

N° answering to
5 and 5 for 0069 read 0169
9 and 21 for 1975 read 1795
7 and 23 for 2711 read 2811
5 and 25 for 4106 read 4206
27 and 44 for 0927 read 0947
19 and 69 for 2542 read 2592
25 and 75 for 1864 read 1894

Page 22. line 13 and 15. for 9214 read 7764
Page 23. line 4. for 30 + 15 read 50 + 25
Page 24. line 12. for prop. log. of 15'. 32" 0.0653 read prop. log. of 15'. 29" 1.0653

Mr. Dunthorne's Tables.

Page 58. against 29° under 53' for 303,5 read 305,5
Page 63. in title (hor. par. ☽) over the last column, for 61' read 62'

In the Table of proportional logarithms.

Log. of 29. 13 for 7996 read 7896
29. 14 for 7994 read 7894
56. 28 for 5735 read 5035
57. 30 for 4656 read 4956

Page 157. line 24 and 25. the two ar. com. fine are each put one line too high.